Maximize Lithium-Ion Battery Power: Your Comprehensive Guide

Melanie .H Gillespie

All rights reserved. Copyright © 2023 Melanie .H Gillespie

Funny helpful tips:

Invest in leadership development; strong leaders drive business growth.

Stay updated with the evolution of digital wallets; they're centralizing financial transactions and enhancing security.

Maximize Lithium-Ion Battery Power: Your Comprehensive Guide : Unlock the Full Potential of Lithium-Ion Batteries: The Ultimate Guide to Boosting Performance and Lasting Power.

Life advices:

Prioritize sleep; it's foundational for health, productivity, and well-being.

Stay agile in business strategies; adaptability ensures resilience.

Introduction

Welcome to the fascinating world of lithium-ion batteries, where innovation and energy meet to power our modern lives. In this book, we embark on an electrifying journey to explore the very heart of these remarkable power sources, uncovering their basics, applications, and the crucial safety precautions associated with them.

Our voyage begins with a deep dive into the fundamentals of lithium-ion batteries. We'll unravel the science behind their operation, demystify their components, and gain a comprehensive understanding of the role they play in a multitude of applications, from emergency power backups to electric cars and even batteries designed for underwater use.

As we navigate the world of lithium-ion batteries, it's vital to shine a light on safety concerns. We'll address the critical issues surrounding these batteries, from thermal runaway and short-circuiting to toxic gas emissions and overcharging. By understanding these potential hazards, you'll be better equipped to use lithium-ion batteries safely and responsibly.

But fear not, for we are committed to making lithium-ion technology safe. Dive into our DIY safety tips, where we provide you with essential guidelines to handle these batteries with care, from gathering your materials to selecting the right battery brand and understanding important cell ratings.

For those daring enough to venture into the world of DIY battery packs, we've got you covered. Learn the art of building your lithium-ion battery pack, from designing the layout and understanding voltage and capacity to selecting the ideal continuous discharge rating (CDR) for your needs. Discover the intricacies of cell matching and get hands-on with assembling the cells, connecting them, and adding a Battery Management System (BMS) to ensure safe and efficient operation.

Choosing the right charger for your custom battery pack is a crucial step, and we'll guide you through the process, explaining charger types and helping you add those finishing touches that make your creation truly remarkable.

But our journey doesn't end with construction. Maintaining your DIY lithium-ion battery pack is equally essential. Learn the ins and outs of charging your pack, explore fast charging options, and discover ways to prolong the life of your lithium-ion batteries. We'll even delve into the fascinating realm of restoring battery packs and offer you practical maintenance tips to keep your power source in prime condition.

As we embark on this electrifying expedition through the world of lithium-ion batteries, we invite you to embrace the knowledge and skills needed to harness their power safely and effectively. Together, we'll unlock the potential that lies within these remarkable energy storage devices, revolutionizing the way we experience and interact with the world of technology.

Contents

Chapter One: Lithium-Ion Battery Basics 1
Chapter Two: Lithium-Ion Battery Applications and Safety Precautions 18
    1. Emergency Power Backups or Uninterrupted Power Supply (UPS) 18
    2. Energy Storage Systems 20
    3. Electric Cars 20
    4. Batteries for Underwater Use 21
    Safety Concerns with Using Lithium-Ion 21
    Thermal Runaway 22
    Short-Circuiting 24
    Toxic Gas Emissions 24
    Overcharging 25
    Reasons for Battery and Cell Failure 25
    1. Bad Cell Design 25
    2. Problems with the Manufacturing Processes 26
    3. Battery Aging 26
    4. Unchecked Operating Conditions 27
    5. External Factors 28
    Making Li-Ion Safe 29
    DIY Safety Tips 31
Chapter Summary 32
Chapter Three: Gathering Your Materials 34
    Cell Ratings 34
    What Is an 18650 Cell? 36
    Selecting a 18650 Battery Brand 37
    Continuous Discharge Rating (CDR) 38

    Capacity 38

Chapter Four: Building Your Lithium-Ion DIY BatteryPack 57

    Battery Pack Layout and Design 59

    Voltage 59

    Capacity 60

    Maximum Continuous Current 61

    Factor of safety = maximum allowable load/actual load 61

    Step 1: The Layout Drawing 62

    Connection Circuits: Series and Parallel Circuits 63

    Cell Matching 65

    Step 2: Assembling the Cells 67

    Step 3: Connecting the Cells 68

    Step 4: Adding the BMS 73

    Step 5: Sealing the Battery Pack 77

    Step 6: Choosing a Charger for Your Battery Pack 80

    Charger Types 81

    Step 7: The Finishing Touches 86

Chapter Summary 87

Chapter Five: Maintaining Your DIY Lithium-IonBattery Pack 88

    Charging Your Pack 88

    Accelerated or Fast Charging 92

    Prolonging Li-Ion Battery Life: How and When to Charge 93

    What Causes Lithium-Ion Batteries to Die 96

    Restoring Battery Packs 100

    Maintenance Tips 102

Chapter Summary 103

Chapter One: Lithium-Ion Battery Basics

Before the development of lithium-ion batteries, most rechargeable battery systems were lead-acid and manganese dioxide-zinc based. However, these systems have major drawbacks, such as slow rechargeability rates and the use of potentially hazardous materials (lead). While lithium-ion batteries are not perfect, they are a great improvement over these previous battery types.

Lithium-ion batteries are made from lithium, which is the lightest elemental metal. What makes lithium ideal for this use is that it has the greatest electrochemical potential and thus provides the largest specific energy per weight. We will expound what this means later when we discuss how batteries work. Lithium-ion batteries offer distinct advantages over other battery technologies. These include:

- High energy density — Lithium-ion batteries have the highest energy density, with lithium-cobalt batteries having the highest energy density. Energy density can be defined in two ways. Firstly, it is defined as gravimetric energy density, which refers to how much energy a battery has in comparison to its weight. This is expressed in Watt-hours/kilogram (W-hr/kg). The second definition is as volumetric energy density, which refers to the energy a battery has in comparison to its volume, expressed in Watt-hours/liter (W-hr/l).

This table shows the energy density comparison between three battery types, namely lithium-ion, nickel-cadmium, and nickel-metal hydride.

Battery type	Ni-MH	Ni-CD	Li-ION
Gravimetric density (W-hr/kg)	55	50	90
Volumetric density (W-hr/l)	180	140	210

As you can see, lithium has the highest energy density of the three most commonly used battery compositions, making lithium-based batteries a common choice in portable devices.

- Low self-discharge rates — The biggest issue faced by many rechargeable batteries is their fast discharge rates. However, Li-ion cells have a much lower discharge rate.

- Low maintenance — Lithium-ion batteries do not require much maintenance to ensure that they perform well, unlike lead-acid cells that can require acid top-ups or Ni-Cad cells that may need to be discharged periodically to ensure they don't exhibit the memory effect.

- Cell voltage — A lithium-ion cell produces on average 3.6 volts (V), which is much higher than the other batteries, as most range between 1.5V to 2V. Since each cell produces more voltage, fewer lithium-ion cells are required in many battery applications. For instance, a phone only needs one cell to power it. This significantly simplifies power management.

- Load characteristics — Load here refers to the current drawn from a battery. For Li-ion batteries, it is comparatively good, as they can provide a constant 3.6V voltage per cell before the last charge is used up.

- It doesn't require priming — Some rechargeable batteries need to be primed before they are charged for the first time. This means that they need to be conditioned to help improve the battery's performance. This is mainly for lead and nickel-based batteries. Lithium-ion batteries come operational and ready for use straight from the factory.

- It has a variety of different battery types — There are many types of lithium-ion cells available today. Since lithium has a lot of advantages, it can be combined with other materials and used for a particular application. Some have high specific energy, which is great for electronic equipment such as phones, while others have a high specific power, which is great for power tools.

Even though lithium-ion batteries have many benefits, they also have several drawbacks, such as:

- The batteries need protection — These cells are not as robust as other rechargeable cells. They need to be protected from overcharging and also from being discharged too far. What this means is that their performance is greatly affected if they are fully discharged too often. It is also important that the current drawn during the charge is maintained within safe limits. This means that the circuit should be designed in a way that doesn't draw too much current from the battery at once and remains within safe operating limits. Luckily, due to technological advancements, there are ways to safeguard against improper charging and discharging by using an integrated circuit.

- Aging — Lithium batteries tend to age quickly compared to other battery technologies. Not only is this degradation time or calendar dependent, but it is also affected by the number of charge-discharge cycles a battery undergoes. This aging happens whether the battery is in use or not.
- Transportation — Since they are used to power many portable electronics, many airlines limit the number of lithium batteries they can transport. Most transportation is done by freight ships.
- Cost — Making a lithium-ion cell is 40% more costly than making a nickel-cadmium cell. This is a major factor to consider in manufacturing as this cost trickles down to the consumer, especially with mass-produced items such as phones.

As with any technology, it is essential to understand its advantages and drawbacks. In doing so, you can understand why we chose to build a DIY lithium-ion battery pack as well as the drawbacks you might face with your pack. With this information, you know what to expect, and you can figure out different ways to overcome the shortcomings and reduce their effects.

Battery Terminologies, Terms, and Definitions

Before we dive into the primary subject matter, we should first look at some of the terms, definitions, and terminologies we will use. You need to understand these terminologies because you will use them later on when we move on to making our DIY battery pack. Some of the terms we will use include:

1. Battery — This is a generic term used to refer to the unit that creates electrical energy from chemical energy. It can consist of two or more cells arranged either in a parallel or series configuration to get the desired voltage or capacity. It also refers to a unit consisting of a singular cell, especially when it has a battery management circuit.
2. Cell — This is the basic electrochemical unit used to create electrical energy from chemical energy and also store electrical energy as chemical energy. A cell consists of an anode, a cathode, an electrolyte, and sometimes a separator. One or more cells make up a battery.

It is important to note this distinction as we move forward because the terms above are used interchangeably.

3. Anode — This is where the oxidation process takes place. During charging, it is the positive electrode and the negative one during discharging.

4. Cathode — This is the electrode where the reduction reaction happens. During charging, it is the negative electrode and the positive one during discharging.

These two elements are known as electrodes in general, and the reversing of polarity occurs due to the changes in the flow of electrons. But to avoid confusion, electrodes are defined for their activity during the discharging process. Hence, the anode is the negative end and the cathode is the positive end. Every cell or battery's voltage is determined by the voltage difference between the cathode and the anode. We will talk about electrodes more when we look at how a battery works.

5. Electrolyte — This refers to the medium that conducts ions to and from the electrodes. It can be a liquid, solid, or even in some cases, such as in lithium polymer cells, a gel medium.
6. Separator — This defines a membrane that is placed in a cell to prevent the electrodes from touching and causing a short circuit. As cells become more compact, it is easy for the electrodes to touch, resulting in a short circuit or even an explosion. It is porous, meaning the electrolyte and ions can pass through it easily, and also electronically non-conductive.
7. Capacity — This refers to the amount of energy a cell or battery can give in a single discharge. It is indicated in amp-hours (mAh, Ah), or watt-hours (Wh).
8. Specific power — This is a battery's gravimetric power density and is expressed in watt/kilogram (W/kg).
9. Specific energy — This is the gravimetric energy stored in a battery, and is expressed in watt-hours per kilogram (Wh/kg).

Earlier, we mentioned specific energy and defined it as the battery's capacity in weight, and it is related to the battery's runtime. A product that requires long running times at moderate power is optimized for specific energy, for instance, mobile phone batteries. Specific power, alternatively, is the battery's capability to deliver high current, and it shows the battery's loading ability. Batteries, such as those used to run power tools with high specific power, have reduced specific energy.

If a battery has a high specific power, it has reduced specific energy and vice versa. Pouring water from a bottle is a great analogy to illustrate this relationship. Think of the water in the bottle as specific energy and the impact of the rate at which you are pouring out the water as specific power.

So, if you pour out the water quickly, you will have high specific power but lower specific energy since the water will run out fast. Alternatively, if you poured it out slowly, the water in the bottle would last longer (specific energy), but the power output would be reduced (specific power).

10. Rated capacity — This is the cell's electrical capacity indicated as ampere-hours. In other words, it is the total charge a fully charged battery can produce in certain discharge conditions.

11. Self-discharge — It refers to the recoverable loss of capacity of a battery. This refers to how a battery or cell loses its charge over time, thus requiring recharging. Even though this is a normal occurrence, several factors such as the technology used and temperature can affect this rate.

12. Charge rate or C-rate — This is the charge or discharge current as a proportion of the battery or cell's rated capacity.

13. Constant current charge — This term is used to refer to a charging process where the current is kept constant regardless of the battery or cell voltage.

14. Constant voltage charge — This also refers to a charging process where the voltage is kept constant over the charging cycle regardless of the current being drawn.

15. Cycle life — This refers to the number of times a battery or cell can be charged and discharged under certain conditions before its capacity falls to 80% of its rated capacity.

16. Cut-off voltage — This refers to the voltage at which the battery management system terminates discharging. Also referred to as the end-of-discharge voltage, it shows the curve followed by a cell when it is discharging.

17. Deep cycle — This refers to a charge-discharge cycle where the discharge occurs until the battery is fully discharged, which is achieved by bringing the battery to its cut-off voltage (80% discharge).

18. Energy density — This refers to the volumetric energy storage density of a cell shown as watt-hour/liter (Wh/l).

19. Power density — This is a battery's volumetric power density shown as (W/l).

20. Trickle charge — This refers to a form of low level charging where the cell is connected continuously or intermittently to a constant power source that keeps it fully charged.

How Batteries Work

Before we can build our own battery pack, you must have a comprehensive understanding of how a cell works. This will help you understand what takes place inside a battery, how it stores power, and also how it produces current. With this information, you will know how to connect your pack to increase its voltage, power, or current output.

The first thing is to understand what a battery is. We already defined a battery as a combination of two or several electrochemical cells. It is these cells that store energy as chemical energy, which is later converted into electrical energy when a circuit is connected, allowing current to flow. A cell consists of two electrodes immersed in an electrolyte. The negative electrode is called the anode, while the positive one is called the cathode. Due to the demand for smaller batteries and the volatile nature of many of them, nowadays, the electrode-electrolyte interface is enclosed in a special container. An element referred to as a separator is placed between the electrodes.

Electricity or current is the movement of electrons through a conductor. When the electrodes are connected to each other through a conductor such as a wire, they form a circuit that allows for the flow of electrons, resulting in a current. The diagram below shows a simple circuit.

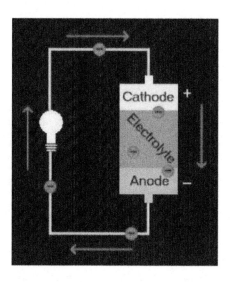

    The chemical reactions that occur in the cell cause the electrons to build up at the anode. This build-up results in an electron imbalance between the electrodes. This electrical imbalance can be resolved since electrons repel each other, so they move to where there are fewer electrons (the cathode). However, the electrolyte prevents them from flowing to the cathode within the battery, so they require a closed circuit to flow through instead. In the above illustration, you can see this flow as the move to the cathode powering the lightbulb along the way. This is a basic description of how electrical potential causes electron flow resulting in current as the electrons flow to the cathode through the circuit.

    These electrochemical processes cause changes to the anode and cathode to stop the supply of electrons, thus limiting the power available in a cell. Recharging a battery reverses the electron flow. At that point, electrons move from the cathode to the anode. This restores them to their original state, and they can provide full power.

    The anode in lithium-ion batteries is made of a porous carbon such as graphite coated in copper. Carbon is preferred because its electron structure makes the material highly conductive. The cathode is made up of lithium metal compounds. In previous iterations, the cathodes were made of lithium alone, but the resulting batteries were too unstable, so people switched to lithium-ion and other lithium compounds. Different electrode materials and electrolyte solutions produce varying chemical reactions that affect how the cell works, how much energy it can store, and the voltage it produces. And although newer battery models had reduced specific energy, the tradeoffs were worth it. The batteries became safer, more stable, and were more easily modified to suit specific needs, such as increased specific energy or increased specific power, and so on.

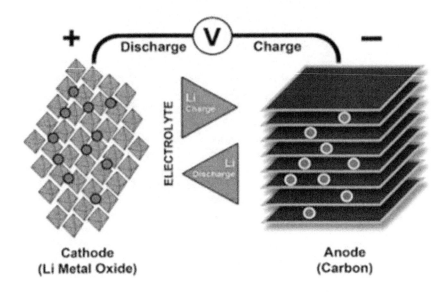

Cathode (Li Metal Oxide) Anode (Carbon)

The key to lithium-ion success is the high cell voltage (3.6V). The improvement of the cathode materials and even the electrolyte offered a boost in energy density.

As we stated earlier, the electrolyte in lithium-ion batteries can be a liquid, gel, or solid (dry polymer). Liquid electrolytes are flammable because they are not aqueous solutions. They are instead solutions of lithium salts mixed in organic solvents. Mixing in carbonates can increase the conductivity capability while increasing the temperature range. Temperature can have significant effects on the performance of a battery. Other slats can also be added to mitigate issues experienced in other batteries, such as gas, as well as boost high-temperature cycling.

For a Li-ion battery to work, a passivation layer must form around the anode. This layer or film is referred to as a solid electrolyte interface (SEI), and it behaves similarly to a separator. It keeps the electrode from touching, stabilizes the system, and increases Li-ion battery lifespan because it causes a capacity reduction. However, when capacity reduction occurs at the cathode, it causes a permanent reduction in cell capacity. To prevent permanent reductions, additives are added to the electrolyte. Since these additives are often consumed during the formation of the SEI, they have become a trade secret among manufacturers.

One known additive is vinylene carbonate (VC). It can increase the cycle life of a Li-ion battery, especially at high temperatures, while keeping resistance low with age and usage. It also keeps the SEI stable without causing electrolyte oxidation at the cathode.

For lithium-ion cells, the SEI film can break down if the battery temperature rises to 75°C-90 °C. This rise in temperature affects different cell types and their charge states differently. If the battery heats on its own, it can lead to thermal runaway or even explosions if the battery is not properly cooled. The flammability of the electrolyte also poses a danger as the temperature rise can also trigger explosions or cause severe burns. The electrolyte may also turn into a solid due to age, thus reducing the battery's performance. Once the liquid electrolyte solidifies, the batteries are dead.

Even though there are many distinct types of batteries, they can all be classified into two broad categories, namely primary or secondary batteries. Each category has its advantages, disadvantages, and applications, although they can often be used interchangeably.

- Primary batteries — These are batteries that are not rechargeable. The process through which they turn chemical energy into electrical energy is irreversible. However, this process can take place many times, depending on the battery type. When all the chemicals in the cell have reacted to create electricity, they cannot be easily restored electrically. Alkaline, silver-oxide, zinc-carbon, and mercury batteries are examples of primary batteries. They are also referred to as single-use cells

- Secondary batteries — Unlike primary cells, these cells can be recharged. This is because the electrochemical process can be reversed using electricity. This electrical energy restores the battery to its original form and composition. The rechargeability factor makes these batteries very popular. They are used to power smartphones, laptops, and even hybrid electrical cars. Examples include Nickel-cadmium, lead-acid, lithium-ion, and more.

Lithium Cell Form Factors

Lithium cell form factors are also known as battery configurations. This term refers to the shape of the battery. In Li-ion batteries, they come in four major configurations, namely coin/button cell, cylindrical, prismatic, and pouch.

Button/Coin Cells

As the name implies, these cells take the shape of a button or a coin. This cell design is favored for its compact design which enables users to power various small devices such as watches, phones, car keys, security wands, medical aids, remote controls, and even small LED flashlights. To increase the voltage of this battery type, you can stack them up or put them in a tube-like configuration. Although they are cheap to make, their capacity is limited, and so they are not used for applications that require more power. The button battery design has a couple of drawbacks, such as the fact that it does not have a safety vent to let out any gases that might be produced in the cell. If you charge it too quickly, the battery can swell, so you have to be very careful not to exceed the recommended 10 to 16 hours charging time. Its small size also makes button cells a choking hazard, so precautionary measures must be taken to keep children from reaching them.

Below is an illustration of a button cell and its main components.

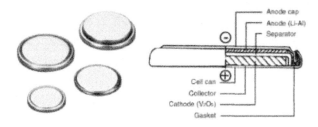

Cylindrical Cells

This is likely the most familiar battery shape design to most of us. You have likely come across this battery form as it is the most widely used packaging style. The cylindrical design has good cycling capabilities, offers a long cell lifespan, and is economical. It is easy to manufacture and mechanically stable since its tubular shape can withstand high pressure. However, the cylindrical battery tends to be heavy and has a low packaging density due to its space cavities.

Many tubular cells have a pressure relief mechanism that uses a membrane that ruptures at high pressures. Still, leakages and dry outs may occur once the membrane breaks. As a result, many cylindrical cells have resealable vents with a spring-loaded valve, which opens to relieve the pressure. Many Li-ion and nickel-based cylindrical batteries have a positive thermal coefficient (PTC) switch, which heats up and becomes resistive when too much current passes through it. It acts as a short circuit protection mechanism that stops current from flowing. But once the PTC cools down it, conductivity returns and current can flow. Other Li-ion cells have a charge interruption device that causes physical and irreversible disconnection of the battery when unsafe pressure builds up.

The following diagrams show the components of a standard cylindrical cell.

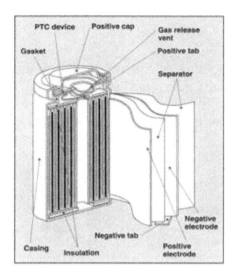

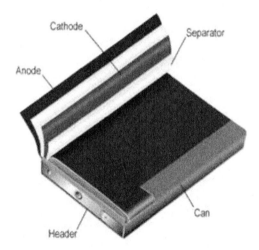

These batteries are used to run power tools, electric bicycles, medical equipment, and so on. Manufacturers use different cell lengths, such as half or three-quarter styles, to create more variation and increase their use. NiCd batteries have the largest variety of cylindrical cell choices, followed by nickel-metal-hydride. However, with lithium-ion, the choices are smaller as the cell chemistry necessitated different formats. The 18650 is the most popular cylindrical Li-ion battery-package available. Other formats include the 20700, 21700, and the 22700 batteries. These formats are relatively new but offer more capacity and other benefits over the 18650 in response to growing consumer needs.

Prismatic Cells

This battery packaging style was first seen in the early 1990s, and it was developed in response to the need for thinner batteries. Prismatic cells are wrapped in thin cases and make optimal use of space inside the cell by using a layered approach. But other designs wind the materials together and flatten them into a prismatic jelly roll. They need to be firmly enclosed to achieve compression. Some swelling can occur due to gases building up, which causes the cell to grow. Such volumetric change is absolutely normal, and it occurs after 500 cycles. However, you should discontinue use if the swelling causes the battery to press against the battery compartment as bulging batteries can cause damage to the equipment and also pose a threat because their safety measures are compromised.

No uniform standards exist, so each manufacturer makes prismatic cells as per their specifications and needs. You can find these batteries mostly in smartphones, tablets, and low-level laptops. They can also be found in large formats packaged in welded aluminum or steel housings to deliver current capacities of 20 to 50 Ah. These large format versions are used to power trains and electric vehicles.

A prismatic cell is shown below.

Pouch Cells

This battery configuration uses conductive foil tabs that are welded to the electrodes and sealed shut instead of the conventional metallic cylinder or glass to metal electrical feedthrough models. This design offers simple, flexible, and lightweight solutions to battery form factors. You can stack them, but remember to make an allowance for the swelling. Smaller pouches can grow eight to ten percent after 500 cycles, while larger cells grow to that size after 5000 cycles. Also, keep sharp objects that might stress or puncture the cells away. Improvements to the design are being made to resolve excess swelling due to gassing. These batteries can deliver high current loads but perform well under light loading and moderate charging conditions. This battery is the most space-efficient, achieving about 90% packaging efficiency, which is the highest among batteries.

You can reduce the weight of a pouch cell significantly by removing the metallic cover, but it needs support and enough space to expand. Most pouch batteries are lithium polymer cells. No standard size exists for the pouch cell, so the dimensions of a pouch cell are left to the discretion of the manufacturer. Applications for this battery type include equipment that needs high current loads such as drones, model airplanes, cars, and other hobby devices. Large format pouch cells in the 40Ah range serve as energy storage systems because fewer cells make the battery simple.

Gassing is a major, unavoidable issue in pouch cells because gasses such as carbon dioxide (CO_2) and carbon monoxide (CO) are produced whenever the battery is used. As the electrolyte decomposes due to usage and age, it releases these gases. Other stressors such as overcharging and overheating encourage gassing as well. If the pouch balloons with normal use, this can be an indication that it is flawed.

Below is a summary:

- Cylindrical cells have high specific energy, are mechanically stable, and can be manufactured automatically. Its design allows for safety features that are not possible in other form factors. It has good cycling, a long lifespan, and its low cost. The downside is that its design does not offer much packaging density. These cells are used in most portable applications.
- Prismatic cells can be jelly-rolled or stacked and are encased in a steel or aluminum case for stability. This design is space-efficient but pricier to make than the cylindrical cell.

- Pouch cells use laminated architecture in a pouch or bag. Pouch cells are light and cost-effective but susceptible to humidity and high temperatures. Stacking them or applying light pressure can prolong the shelf life and prevent delamination. But don't forget that swelling can occur. Large format pouches are great for light loads and moderate charging, attributes that are growing this battery design's popularity.

Types of Lithium Cells

Lithium-ion cells refer to a group of cells that share similarities but have different chemical makeups. As we stated before, lithium metal, the battery's active ingredient, is unstable, but by combining it with different compounds, its weaknesses are mitigated, and other features are added too. These combinations give different cells various features such as boosting power, current or energy output, performance, safety, and reducing costs. Let's look at some of these batteries.

1. Lithium Manganese Oxide (LiMn2O4)

The cathode for this battery is made from lithium-manganese. It has great thermal stability, which improves the battery's safety, promotes electron flow within the electrolyte, and decreases internal electrical resistance, which can lead to loss of power over time. It has a high discharge and recharging rates due to the lithium-manganese cathode's material and design. The downside is that this battery has a lower capacity and a shorter lifespan.

With moderate heat build-up, it has the capability of discharging current of 20-30A. The structure of the lithium manganese oxide cell also gives it the capability to produce high current pulses of up to 50A. Although, you should be careful because this can cause the cell to heat up and raise its temperature, which should not exceed $80^{\circ}C$. Lithium-manganese oxide cells are mainly used in medical instruments and power tools, as well as in electric and hybrid vehicles. The battery's design allows for it to be maximized for optimal longevity, high capacity, or maximum load current. Lithium-manganese oxide cells come in various high-capacity versions of 1,500mAh, long-life versions of the 18650 cell, and some with a moderate capacity of only 1,100mAh.

To prolong its lifespan and improve specific energy, many lithium manganese batteries are blended with lithium nickel-manganese-cobalt oxide (NMC). The resulting blend gives the battery features that make it suited for use in electric cars such as in the BMW i3, Nissan Leaf or the Chevy Volt. This is because the battery has more electrical capacity, which boosts how long the it can last and thus the car's driving range. It can also give a big current boost to accelerate the car.

2. Lithium Cobalt Oxide (LiCoO2)

The lithium cobalt oxide battery is one of the most common lithium batteries available. It has a carbon anode of primarily graphite and a lithium cobalt oxide cathode. The cobalt and oxygen elements bond forming layers of cobalt oxide layers which are separated by lithium sheets. This layer allows the cobalt ion to lose or gain an electron when charging or discharging. The lithium cobalt battery was the first lithium-ion batteries to be made, and it has the highest energy density that's why it is used in our smartphones, laptops and even in digital cameras.

A drawback to these batteries is that they have thermal instability because the anode can overheat, and at these high temperatures, the cobalt oxide cathode can decompose, releasing oxygen. The combination of the oxygen, flammable electrolyte chemicals, and heat can result in a fire or explosion, which can greatly affect the safety of the battery. To prevent such accidents, the battery has an inbuilt overheating protection system that also prevents the battery from completely discharging, which is also bad. These protection systems, such as venting caps, not only improve safety but can also prevent overcharging, which can have severe consequences.

3. Lithium Iron Phosphate (LiFePO4)

This battery has a high discharge rate because the iron phosphate cathode can withstand high temperatures, which means that the battery has good thermal stability and thus improved safety. This makes it ideal to use as an electric car battery, to power various power tools, or even to store energy in power stations. It has a long cycle of life, which means it can be charged and discharged many times. However, its energy density is lower than lithium cobalt, though it has a higher discharge rate.

Its makeup is similar to the lithium cobalt battery, except that the lithium cobalt oxide is replaced with lithium iron phosphate, which is more stable. This compound's properties offer good electrochemical performance with low resistance, which makes it a useful nanoscale phosphate cathode material. Additionally, the battery has a high current rating, enhanced safety, good thermal stability, and a long life cycle. In comparison to other lithium-ion batteries, the lithium iron phosphate has more tolerance to full charge conditions because no lithium or iron ions remain in the cathode once the cell is fully charged.

4. Lithium Nickel-Manganese-Cobalt Oxide (LiNiMnCoO2 or NMC)

Incorporation of nickel and cobalt into a battery's chemical composition can cause significant changes. Nickel offers high specific energy, and the cobalt-manganese compound provides a stable cathode structure. This gives the battery lowered internal resistance, higher charging rates, good stability, and increased safety. The cathode is one part nickel, one part manganese, and one part cobalt; however, this ratio can vary depending on the manufacturer's specifications. Many lithium manganese cells are combined with a lithium nickel-manganese-cobalt oxide battery to make a battery that can provide high energy but still last long.

5. Lithium Nickel Cobalt Aluminum Oxide (LiNiCoAlO2)

This type of battery is known for its long lifespan, high specific energy, and high specific power. It also has the longest cycle life in comparison to other lithium batteries in the market. Its main disadvantages, however, are the marginal safety and high cost of production. Nonetheless, more research is going into making these batteries safer and more cost-effective. Other improvements are aimed at increasing the power of its cells and making it competitive with other batteries because they are a bit pricier than other lithium-ion batteries available.

6. Lithium Titanate (Li2TiO3)

This type of battery is used in electrical power trains, solar-powered street lights, and UPS. The battery replaced the graphite in the anode of a typical lithium-ion battery, with lithium titanate. The upside of this battery is that it can be charged fast and delivers a high discharge current, which is ten times the rated capacity. Its cycle count is also higher than regular lithium-ion batteries, and it has excellent low temperature discharge even at -30°C. The only disadvantage of this battery is its high price tag since titanium is expensive.

7. Lithium Polymer

This battery replaces the liquid electrolyte with a solid one, which not only improves its safety but also makes it lighter. Since the polymer itself is thin, it allows for great flexibility both in shape and design. What this means is that it doesn't require a rigid case, so it can be very compact. The polymer electrolyte is a gel-like non-conductive material that still permits ion exchange. In earlier designs, the polymer was like a thin plastic film that separated the electrodes, but it was found that the film was unable to allow ion exchange unless heated to approximately 60°C. Adding a gelled electrolyte, which is a micro-porous electrolyte, to replace the traditional separator, solved this problem.

A lithium polymer battery is wrapped in a foil-like case and uses laminated sheets that do not require compression as a means of pressing the electrode together. The packaging is ideal as it reduces the weight by more than 20% in comparison to the rigid case found in other lithium-ion batteries. Lithium polymer cells can easily be shaped to fit into any tablet, phone, or even made to be as slim as a credit card.

Many people wonder what the difference is between a lithium polymer and a lithium-ion. Though they both use identical electrodes, the physical state of the electrolyte differentiates between them. Many Li-ion batteries use a liquid, whereas the lithium polymer uses a gel medium. As such, the lithium polymer battery has a higher specific energy and is thinner than a normal Li-ion battery. Remember the pouch/bag cell? That is an example of a lithium polymer battery. Because it is a pouch-style cell, it still has the same drawbacks. There is gas build-up during charging that can cause the lithium polymer cell to expand. The foil packaging is also less durable than the cylindrical package. However, it does not require a dedicated charger.

Unlike other lithium batteries, a lithium polymer battery can use any cathode combination; what differs is the electrolyte. It is also available in different chemical and structural combinations, each offering different features but often having a trade-off between efficiency, cost of production, and safety.

Chapter Summary

We covered quite a lot about batteries in this chapter. Below is a summary of the essentials:

- We defined what a battery is, the difference between a cell and a battery, and explained how it works.
- We discussed the advantages and disadvantages of lithium-ion.
- We also looked at some of the terminologies, terms, and definitions we will use in the book in relation to batteries.
- We also talked about the different types of battery packaging styles. These are also referred to as battery form factors, and they include the cylindrical, prismatic, pouch, and button cell configurations.
- And finally, we discussed the different types of lithium batteries available. In

the next chapter, we will take a look at some of the applications of lithium-ion batteries and the safety precautions you should observe when handling them.

Chapter Two: Lithium-Ion Battery Applications and Safety Precautions

Real-world applications for lithium-ion batteries go further than just powering your phone. From powering lifesaving medical equipment to proprietary technology, lithium-ion batteries keep the essentials and comforts of everyday modern life running safely and reliably. As they are available in various forms and sizes, Li-ion batteries are the perfect power option regardless of the size of the system. These cells provide energy solutions that cut across the board from energy storage to providing portable power solutions. Let's discuss some of the top uses for rechargeable Li-ion batteries.

1. Emergency Power Backups or Uninterrupted Power Supply (UPS)

A sudden loss of power or instability is the primary cause of damage to electrical equipment. And even though you might use surge protectors to safeguard your equipment from power surges, you are still unprotected from brownouts, blackouts, drops in voltages, and other power supply problems. To protect your sensitive electrical equipment against electricity supply issues, you require an emergency power backup. These UPS units function like surge protectors but on a larger scale, and they also provide a buffer against electricity supply interruptions. This buffer can last from a few minutes to an hour, depending on the backup's size and capacity.

Think of a UPS in the following way. Say you are working on your laptop at home to finish some reports, and you have your laptop charging via a surge protection power strip. But suddenly, a blackout occurs. But though you have no power now, your work is uninterrupted because your laptop switched to battery power seamlessly, keeping the flow of electricity uninterrupted. Now you have enough time to finish your work, save or send it, and safely power down your computer.

Most electrical equipment lack the built-in batteries that laptops have. If you were using your desktop when the power outage happened, you would have had to stop writing your report immediately, unsure whether or not your work was lost. The sudden unsystematic shutdown also imposes a great strain on the machine. And it can lead to failure. However, a UPS can give you some time to save your work and safely shut down your desktop computer or put it in hibernation mode. If the power comes back on while the UPS still has enough power, then you can continue with your work uninterrupted. The great thing about Li-ion batteries is that they come with software that detects when they switch to battery power and shut down automatically even when you are not present.

These backup batteries are usually Lithium-Ferro-Phosphate cells. This is because they have long shelf lives, which is an added advantage. They also vary in capacity from small, lightweight ones used to power laptops to large units used in data centers to protect servers from power supply issues.

There are three types of UPSes, namely the standalone units, line-interactive units, and the online UPS units. The standalone UPS is the most common type. It charges its battery and waits for power outages or fluctuations to happen. The switch to battery power takes about 20 to 100 milliseconds, which is well within the threshold of many electronic devices. The online UPS is the costliest of these three units since it requires more circuitry than the other two. Rather than jumping into action when the power goes out or in case there are fluctuations, it continually filters the wall power through its battery. Since the electrical equipment runs on the unit's battery, which is topped off by the wall power, there are no power or voltage issues. Lastly, the line-interactive UPS unit is similar to the standalone version, except that it has a special transformer, which makes it better equipped at handling power sags and brownouts where the lights become dim, but a power outage does not occur.

2. Energy Storage Systems

Battery energy storage systems (BESS) are a subset of energy storage systems. This term is used to define a system that stores energy using thermal, electromechanical, or electrochemical means. All energy storage systems capture and store energy for later use. Examples of such storage systems include pumped hydro and compressed air storage, flywheels, and BESS, which are also an example of electrochemical storage. These storage systems are used for intermittent power sources such as solar, wind, or tidal power so that they can balance power production and use, thus reducing the strain of peak electric power demand. Lithium-ion batteries are preferred in these systems because they offer high energy density, and they can be charged and discharged frequently in their lifetimes. Their cost of production is going down, making them readily available. These intermittent power sources generate low resistance charging, which is very compatible with Li-ion cells. The fast charge feature allows for quick energy buildup within a short period allowing for the maximization of either solar, tidal, or wind power energy potential every day.

3. Electric Cars

As society continues to look for ways to reduce our cars' dependence on fossil fuels, fully electric and hybrid electric vehicles have risen in popularity. Many batteries are used to power electric vehicles, and deciding which battery is best depends on its energy storage efficiency, production costs, constructive characteristics, safety, and lifespan. Lithium-ion batteries are the most utilized technology in electric cars. This is, in part, thanks to their high energy density and increased power per mass cell unit. This has allowed for flexibility in the design allowing for the development of powerful batteries with reduced weight and size at fair prices.

Lithium batteries have high specific energy and show increased power in comparison to Ni-MH and other batteries. They also don't have the memory effect, which is the gradual loss of maximum energy capacity due to repeated charging without being totally discharged first; this results in an increased lifespan. However, high operational temperatures produced due to the chemical reactions taking place in the battery can affect performance and lifespan and undermine the battery's safety.

4. Batteries for Underwater Use

Water and electricity have never mixed well, so it is only natural that there would be numerous problems faced when trying to find ways to power devices that run on and in water such as yachts, underwater scuba diving gear, trolling boats, and others. With Li-ion batteries, you can focus on what you are doing, whether it's having fun or exploring, without worrying about how exposure to water will affect your equipment.

Safety Concerns with Using Lithium-Ion

Lithium-based batteries provide high energy density and low self-discharge rates alongside other benefits that put them ahead of other rechargeable batteries. They are actively promoted as viable energy solutions and are used to power most of the devices we use. With these batteries being used by millions of people, it is important to understand the risks behind this battery technology. For all the advantages lithium-ion batteries have, there remain some inherent dangers to using them. From exploding batteries to emitting toxic gases, it is becoming clear that even though lithium-ion batteries have made great strides in giving us more flexibility with our power options, they are still far from perfect.

Incidents involving lithium battery fires have been increasingly headlined in recent years. From Samsung recalling the Galaxy Note 7 because the batteries seemed to explode during charging to Dell recalling laptop models due to similar concerns, it is clear that there is a serious problem. While these batteries have the potential of producing large amounts of energy, there is also the risk of explosions and fires associated with them.

Lithium batteries can fail for several reasons, such as overcharging, damage, overheating, short-circuiting, internal failures, and production deficiencies. Let's examine some of the issues affecting lithium-ion batteries.

Thermal Runaway

Li-ion cells can fail due to overheating in a process referred to as thermal runaway. This is a reaction that happens inside the cell, causing the temperature and pressure to rise faster than it can be dissipated. Once the cell is in thermal runaway, it generates enough heat to affect its surroundings. If the cell is joined to other cells, whether in a battery or a battery pack, one overheating cell can quickly cause its neighboring cells to overheat as well. This generates a fire that can spread quickly as each cell bursts releasing its contents. To prevent this chain reaction,, most battery cells are separated with dividers or packed individually and insulated. This insulation helps prevent the heat from spreading.

Remember that electrolytes are flammable, so this fire should not be treated as a normal fire. It is different and uniquely dangerous and thus requires special measures to stop and extinguish it. Even after the fire is put out, keep watch over it as some batteries can reignite several hours later. Take lithium batteries with cobalt cathodes; their temperatures should never rise above 130°C. If their temperatures become higher than this upper limit, the batteries become unstable, leading to overheating and release of toxic gases.

When charging a battery, ensure that the charging current, voltage, and temperate are regulated to keep the battery from heating up. If it heats up, the internal resistance lowers, allowing more current into the cell, raising the temperatures more.

There are several ways to help prevent thermal runaway. These include:

- Separating the cells or batteries by using a material that does not conduct heat. Alternatively, you can insulate the batteries to prevent the heat from moving to other cells or it coming from other cells.
- Keeping the overall temperature of the battery low to maintain pressure and prevent high temperatures. You can do this by keeping your batteries out of the sun and hot environments.
- Opt for batteries with a manganese spinel as it has higher thermal stability even though its energy density has gone down. It can take temperatures of up to 250°C.
- Lithium-ion battery manufacturers have also placed fail-safes to help make Li-ion cells safe and reliable once more. These layers of protection include decreasing the quantity of the active ingredient to strike a balance between energy density and safety. Consider adding safety measures such as the venting cap to the cell or including an electronic protection unit called a protection circuit board to the battery pack.

It is crucial to note that these safety measures only work with external interference, such as a faulty charger or an electric short circuit. Under most circumstances, Li-ion batteries simply power down when a short occurs, but if the cell is contaminated with metal particles, this anomaly would not be detected. This means that the protection unit will not stop the cell from disintegrating once thermal runaway has begun as the process cannot be stopped.

Short-Circuiting

Sometimes during the manufacturing process, microscopic particles of metal can come into contact with some of the cell components resulting in a short circuit. A mild short can only cause a higher self-discharge rate with mild heating because the energy being discharged is low. However, if the metallic particles were to gather in a single spot, they would cause a major short circuit, which would cause a higher current to flow. This can result in the generation of higher temperatures and a buildup of pressure (thermal runaway). Even though manufacturers try to reduce the presence of these metallic particles in the batteries, it can be increasingly hard to make this assurance when using complex assembly techniques. Eliminating metallic dust is near impossible.

Toxic Gas Emissions

A fully charged Li-ion battery will release more toxic gases than a half-charged battery. This is due to the reactions that take place in the battery during the charging process. Some of the carbon atoms from the anode bond with oxygen molecules released by the cathode to form carbon dioxide or carbon monoxide gas. The gases produced can vary in concentration and type depending on the cell's capacity to release charge.

Take carbon monoxide, for instance. It can quickly cause a lot of harm if it leaks in a small, constricted space such as a bedroom, car, or train compartment. Carbon monoxide is produced when Li-ion batteries are exposed to high temperatures, such as when batteries overheat due to charging or being left out in the sun or if they are damaged in some way. By identifying the gases and why they are emitted, manufacturers can find better ways to mitigate the gas emissions and protect the masses since most products are powered by Li-ion cells.

Overcharging

The Li-ion battery's adaptability to any charger is both a blessing and a curse. Many of the fire incidents reported seemed to have occured when the batteries were charging. The use of an ordinary charging port may be part of the problem since not all USB ports are created equal. Charging ports have varying currents and voltage loads, which can cause devices such as hoverboards, e-cigarettes, or phones to overcharge. Ensure that you use the manufacturer's power cord and adapter to charge your devices to prevent overcharging the battery.

Reasons for Battery and Cell Failure

Since these batteries have numerous applications, it is important to know whether or not your batteries are reliable. Knowing the common reasons for battery failure can help you identify faulty cells and avoid using them. This matters because we are building a DIY battery pack, and how you design the layout of the cells and use the battery pack will be vital in mitigating any issues that might arise. It will help you come up with safety strategies to mitigate catastrophes such as explosions, short circuits, and thermal runaway.

1. Bad Cell Design

This is the most practical place to start and also the one you are least able to influence. Bad cell design can cause faults such as weakness in the cell's mechanical form, inadequate pressure seals and vents, and use of substandard materials to save on costs, and improperly specifying tolerances. Since many Li-ion users don't know about the chemistry or engineering used in making these batteries, there is little we can do to assure ourselves of the quality of the cells other than taking the manufacturer's word.

2. Problems with the Manufacturing Processes

Some of these issues include:

- Manual production methods — It can be tough to achieve precision, repeatability, and maintain a contamination-free zone while using manual assembly lines. Potential problems include contamination, leaks, short circuits, and unreliable connections.
- Damage can occur to the battery components pre-assembly.
- Voids can occur due to inadequate compression, causing a reduction in cell capacity, increased impedance, and hamper heat dissipation efforts. Impedance is the active resistance a component has to alternating current. It originates from the combined outcomes of ohmic resistance and reactance.
- Use of unapproved materials during production — This can lead to mechanical weaknesses resulting in leaks, cracking, distortion, or even splitting.
- Poor control over the morphology of the electrodes, which requires the particles to be really small and uniform to achieve the specified cell's capability.
- The quality of the welding or sealing methods and materials used. If batteries are not sealed properly, they can malfunction, providing unstable connections and leading to localized heat buildup. Poor sealing leaves the cell susceptible to leakages, water ingress, and corrosion, creating potential safety issues.
- Damaging the separator during the welding process.

3. Battery Aging

A battery's performance gradually decreases with time due to physical changes and unwanted reactions. This process is irreversible and eventually causes the battery to fail. Here are some examples of problems caused by aging.

- Passivation — This effect is caused by the gradual thickening of the resistive layer that builds up around the electrodes hampering the chemical reactions required to charge and discharge a battery. It also increases impedance while reducing the quality of the active ingredients in the cell. This layer is known as the solid electrolyte interface (SEI), which regulates the reaction happening at the cathode and restricts current flow. It forms during the initial charging, and it is an essential part of the cell formation process.

- Corrosion consumes the active ingredients in the battery, causing impedance, capacity reduction, evaporation of battery chemicals, and gassing.

- Physical changes to the makeup of the battery's components. This can include crystallization of the electrode, reducing the area where active reactions can occur, and also dendritic growth. This is another form of crystallization where the crystals form tree-like structures on the electrodes. It can result in increased self-discharge rates, and if they grow too much, they can damage the separator causing a short circuit.

- Cracking of electrodes or electrolytes in cases where solid electrolytes are used.

All of these aging processes are accelerated by high temperatures, whether internal or external.

4. Unchecked Operating Conditions

Batteries are not immune to failure due to misuse and abuse. This abuse or misuse can result in high battery temperatures, which can be caused by:

- Bad application design, such as using the wrong battery to power a device.
- Unsuitable charging conditions such as using the wrong charger or a faulty one.
- Overcharging
- High ambient temperatures causing a rise in cell temperature.
- Physical abuse and damage such as dropping, immersion in fluids, being in extreme temperatures either too hot or freezing temperatures, crushing or puncturing the battery. Any of these things can cause catastrophic failure.
- Storing cells in high or low-temperature environments.

5. External Factors

Battery failure may also occur due to the malfunction or failure of an external system installed to increase the safety of the battery by mitigating some of the risks. These include circuit interruption; sensor, fan, BMS, or charger failures; loss of cooling fluids; and BMS malfunctions resulting in a loss of communication with other battery protection systems. Performing a diagnostic on the battery can help identify the cause of battery failure.

Below is a summary of the reasons why cells fail:
- Separator damage
- Overheating
- Thermal runaway

- An increase in internal pressure
- Gas buildup
- Swelling of the cell
- Increased self-discharge rate
- Capacity reduction
- Electrode plating
- Electrolyte breakdown or loss
- Changes in the molecular and physical makeup of the cell components
- Exhaustion of the active ingredients in the cell
- A rise in internal impedance

Unfortunately, most of these changes are irreversible; however, by understanding them, you can better handle and prepare your battery packs to mitigate these effects and make your battery last longer.

Making Li-Ion Safe

Since lithium batteries are everywhere, the chances that something might go wrong are quite high. Even though the use of Li-ion materials have addressed some of the issues that faced lithium batteries, such as the unstable nature of lithium as an active ingredient, there are still a few issues. Here are some of the measures you can take to ensure that you are safe.

1. Keep them away from open flames or any flammable environments as these can cause internal temperatures and pressure to rise, causing thermal runaway. Ensure that the batteries are always at room temperature.

2. If you have spare batteries that you are using for your devices, it is advised that you store each one separately to keep the terminals from touching. You can place them in their own cases, in a bag, or tape the terminals. This will prevent any short-circuiting from happening.

3. Ensure that your battery is not damaged in any way. This is because they release toxic materials that can harm you or your device. Regularly inspect it for punctures, dents or tears and avoid using damaged cells.

4. Keep your extra batteries in a secure, dry, airy, and cool place. The temperature should be below 25°C. Keep them away from direct sunlight as it can cause the battery to heat up.

5. Another issue with the temperature that you should be aware of is cold temperature charging. Consumer-grade Li-ion batteries cannot be charged when the temperature is 0°C. Even though they appear to be charging normally, lithium metal plating occurs at this temperature. It is permanent and cannot be removed. If lithium plating occurs often, it can compromise the battery, making it more susceptible to failure caused by other factors.

6. Li-ion batteries should be kept separate from other battery chemistries and components. So if you want to store your lithium batteries, keep them in their original casing. For recycling, tape the ends before you place them in the recycle bins.

7. Clearly label boxes or rooms where lithium batteries are stored. This will help in notifying everyone that these storage areas are out of bounds. Ensure that you have a Class D fire extinguisher or one meant to put out metallic

fires nearby when making your DIY battery pack or placed in energy storage system rooms.

8. Don't use unidentified Li-ion batteries from sources you don't know or are unsure of. Non-brand-named lithium batteries are dangerous because you cannot be sure of their capacity or if they have the safety measures mentioned above. These batteries are made in response to the growing need for cheaper replacement batteries. Manufacturers are now securing their batteries with a secret code that only matches with the device they were made for. This is especially important with laptop batteries.

Lithium is considered hazardous waste, and so its disposal should only be handled by well-trained and certified personnel. Once you are sure that your battery is dead, tape the terminals and dispose of it at a recycling plant.

DIY Safety Tips

When doing a DIY project, there are a few precautionary measures you must take. Lithium is a volatile metal that can react with water or moisture in the air. It is flammable and should be kept in a container that will not let moisture in. Here are a few safety tips to observe while building our battery pack.

1. Work in a well-ventilated area, especially when you are using a Li-ion.

2. Ensure that the batteries don't come into contact with any water or moisture.

3. Keep metallic objects away from the battery terminals as it can cause a short circuit. Also, keep the battery tips from touching.

4. Keep the batteries in a cool, moisture-free place away from direct sunlight and other heat sources.//-//
5. Don't stack heavy objects over the batteries as they can cause physical damage to the battery. However, if you are stacking them to increase the battery's capacity and stability, remember to use a separator and not to stack too many on top of each other.
6. Wear the right attire for the task. This will prevent spills or other possible contaminants from getting in the setup.
7. Use power tools safely to prevent injuring yourself or others.
8. Be very keen about your surroundings and have a first aid kit on hand.
9. If the contents of the cell come into contact with your skin, eyes, ears, and nose, it may cause irritation or burns. If this happens, flush the area with warm water and seek medical help immediately.

In everything that we do, safety is an essential factor to consider. That's why you must understand how to handle lithium-ion batteries, the risks involved with using them, and how you can keep yourself safe. And since we are working on a DIY project, we also included some safety tips you should observe while working on your DIY battery pack.

Chapter Summary

In this chapter we covered:

- Some of the applications of Li-ion batteries.
- The risks involved with their use, such as fire, explosion, and production of toxic gases.

- We also looked at some of the reasons why batteries fail, the steps you can take to make using Li-ion safer, and some DIY safety tips to observe.

In the next chapter, we will gather the materials for our DIY battery pack as well as take an in-depth look at some of the components of the battery pack.

Chapter Three: Gathering Your Materials

As we have seen, lithium-ion batteries are pretty great. Not only are they safer than single-use batteries, but they are also highly reliable and can provide enough power for any need you might have. So logically, you'd want to learn how to make your own Li-ion battery pack. Having understood all the above concepts, definitions, and safety tips, we can move on to making our own DIY battery pack. For our project, we will use the 18650 Li-ion cells to build our battery pack. We chose these cells because they are cheap and readily available and can easily be configured to increase power or voltage. Do note, some of these configurations cannot be applied to other cells, such as prismatic or pouch cells.

Cell Ratings

Before we look at the 18650 cells, there are a few battery specifications you must understand, namely the battery energy capacity, cycle life charge, voltage, and current. You have probably seen a couple of battery cells, such as AA, AAA, and D cells. However, as users, we do not pay much attention to what is written on the battery's casing beyond the purpose, size, and the ability of the pack to power our devices. But you must understand these battery ratings as they can help you decipher a cell's specification and also help you in matching what your DIY pack will provide with what is needed. These rating specifications include C-rate or E-rate, capacity, the maximum number of cycles, maximum charge rate, and maximum discharge rate.

- C- and E- rate — A battery's discharge current is often expressed as a C- or E- rate to normalize against battery capacity, which often varies between batteries. A C-rate is a measurement of a battery's discharge rate relative to its maximum capacity. A 1C rating means that the discharge current will exhaust the entire battery in 1 hour. For a 20Ah battery, this equates to a discharge current of 20Amps. If it has a 5C rating, then the battery's discharge current would be 100 Amps, and a C/2 rate would be 10 Amps. An E-rate also describes a cell's discharge power. A 1E rating is equal to a 1C rating. C rate does not change based on the battery type, meaning that two identical cells with different capacities can have identical C rates. It is also inversely proportional to the capacity of the battery. So an increase in capacity would mean a decrease in the C rate. If a cell is rated 2500mAh, it means that it can deliver a current of 2.5A (1C) for an hour before the voltage drops to a specified point. The same cell will provide 5A (2C) for 30 minutes, 10A (4C) for 15 minutes, and so on.

- Maximum charge rate — This is the highest rate at which you can charge a cell. For many cells in the market, including the 18650, this rate is 0.5C.

- Maximum continuous discharge current — This is the maximum current at which the battery can be discharged continuously without damaging the cell.

- Maximum 30-sec discharge pulse current — This is the maximum current at which the battery can be discharged in pulses lasting no longer than 30 seconds.

- These two rates are often set by the manufacturer to prevent discharging the cell excessively as it would damage the battery or decrease its capacity.

- The maximum number of cycles — This defines the number of times a lithium-ion battery can be charged and discharged before its capacity dips, and it is considered dead. For lithium-ion cells, this number can be greater than 1000.

What Is an 18650 Cell?

An 18650 cell is a rechargeable Li-ion cell with a voltage of 3.7V and a capacity ranging from 1800mAh to 3500mAh. There are two types of 18650 cells: protected and unprotected. Protected 18650 cells have an electric circuit embedded in the cell's packaging or the battery casing, which protects the cell from overcharging, over-discharging, heating, drawing too much current, and short-circuiting. Unprotected cells lack this protection circuitry, thus protected 18650 cells are safer. Even though the unprotected cells are cheaper, it is not advised to utilize them. However, this does not mean that unprotected cells cannot be used to make a battery pack. They can be used in a system where the drawing and charging of the battery are controlled and monitored externally. In general, unprotected cells will have a button top, but always check the battery specs to be sure.

These batteries are generally referred to as 18650 because their dimensions are 18 mm by 65 mm. Other sizes include 14500, 20700, 21700, and 26650.

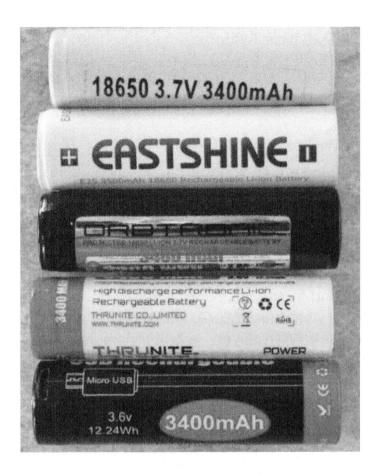

This image shows examples of different 18650 cells.

Selecting a 18650 Battery

Brand

When selecting your 18650 battery cells, an important factor to consider is the battery brand. While this might seem a bit strange, understand that not all brands are made equal. Many times you will find that many of the 185650 cells in the market are misleadingly rated. This is because of the popularity of lithium-ion battery technology. The goal of any manufacturer is to sell as many cells as possible, so some companies, especially Chinese re-wrapping ones, are known to exaggerate their ratings. Nonetheless, this practice is unlawful and dangerous. Luckily, due to the popularity and volatile nature of lithium, there are watchdogs that monitor the battery industry, ensuring that the battery products in the market are at par.

Continuous Discharge Rating (CDR)

When selecting a cell, you need to know how much current the device or equipment you want to power needs. If you pick a cell with a lower current rating than you need, the battery will overheat because it is working beyond its ability. A cell's CDR indicates the maximum current that can be drawn from the battery continuously and safely as per the manufacturer's testing without damaging the cell. Going beyond these limits can severely damage the battery and increase the risk of battery failure, which poses an even greater danger to the user. Alternatively, a battery's pulse rating shows the maximum current a cell can provide while being discharged for a short period without damaging or reducing the capacity of the cell. This rating, however, is not used as a standard because there are too many varying factors in play such as the pulse length, rest time between pulses, battery temperature to conclusively compare two cells. Instead, the CDR rating is used.

Capacity

The next factor to consider is capacity. We all want a cell with the highest capacity. A battery's capacity tells you the amount of current you can draw from the battery and for how many hours before it discharges. It is measured in milliampere-hours (mAh), and it comes at the cost of current rating (CDR). Since a cell can only fit so much material, when picking one, you have to choose between a high capacity or a high current cell. You cannot have a cell with both since one feature is a tradeoff for the other. For instance, the LG HB6 18650 cell has a 30A CDR but a capacity of 1500mAh, while the Panasonic NCR18650B had a capacity of 3400mAh but a CDR of 4.9A. There are, however, batteries that try to balance the two, such as the Samsung 25R, which has a capacity of 2500mAh and a CDR of 20A, or the LG HG2, which has a CDR of 20A and a 3000mAh capacity.

Here is a listing of several 18650 cells available in the market. It can help you choose a suitable cell with the capacity and CDR rating you need for your battery pack.

Brand	Model	Capacity (mAh)	Discharge rating (CDR)
MXJO	IMR3500	3500mAh	11A
MXJO	IMR3000	3000mAh	20A
LG	MJ1	3500mAh	10A
LG	MH1	3200mAh	10A
LG	HG2	3000mAh	20A
LG	HE2 & HE4	2500mAh	20A
LG	HD2	2000mAh	25A
LG	HB2 & HB6	1500mAh	30A
Sanyo	NCR18650GA	3500mAh	10A
Sanyo	UR18650NSX	2600mAh	20A
Samsung	30Q	3000mAh	15A
Samsung	26F	2600mAh	15A
Samsung	25R	2500mAh	20A
Samsung	35E	3500mAh	08A
Panasonic	NCR18650B	3400mAh	05A
Panasonic	NCR18650PF	2900mAh	10A
Sony	VTC6	3000mAh	15A
Sony	VTC5A	2600mAh	25A

Sony	VTC4	2100mAh	30A
Imren	IMR3200	3200mAh	15A
Imren	IMR3000	3000mAh	20A
Efest	IMR18650V1	3500mAh	10A
Efest	IMR18650V1	3000mAh	20A
Vapcell	INR3000	3000mAh	20A
Vapcell	INR2000	2000mAh	38A

Voltage

A battery's voltage refers to the difference in electrical potential between its electrodes, also referred to as electromotive force. The greater the voltage, the greater the flow of current. You will want a battery that maintains its voltage through the discharge cycle, as a dip in voltage results in a sag in power output. The efficiency of a battery is a critical factor, especially in larger battery systems. This efficiency is measured by determining the cell's coulombic efficiency, which describes the efficiency by which electrons are transferred in cells. It is a ratio of the total charge put into the cell to the total charge drawn from the cell in a full cycle.

Li-ion has the highest coulombic efficiency of all battery types, exceeding 99%. This is only achieved when it is charged at moderate current and in cool temperatures. The fast charging feature boasted by many of these rechargeable batteries lowers the coulombic efficiency due to the losses incurred because of charge acceptance and heat generated. You would think that slowly charging the battery would work; however self-discharge comes into play. Different battery systems will have differing coulombic efficiency values depending on the charge rate, battery age, and temperatures. This should be kept in mind when you are sourcing your batteries.

Operating Temperature

You may have noted that temperature, whether internal or external, affects how a battery works, its efficiency, and its capacity. High temperatures can cause problems such as premature battery aging, explosions, venting, and can easily damage the equipment. If your battery constantly gets hot when you connect it to a circuit, it may be an indicator that the circuit is too taxing or the battery is faulty.

Terminal End Type

The last thing you might want to consider is whether you want a cell with a flat or button positive terminal top. This matters because this little protrusion can increase the battery's size and its ability to fit. Some devices are specifically designed to be used with flat top cells, and others, with the button top. The flat-top cells can be shorter, so ensure that you get the cell with the right terminal top.

18560 batteries come in a wide variety of chemical mixtures. You must note this as some combinations are more volatile than others, although each has its benefits and drawbacks.

Caring for Your 18650 Battery

Utmost care should be taken when using, handling, or storing Li-ion batteries. Please refer to the previous chapter on Li-ion safety for these tips. Before we get to care instructions, however, here are some signs that can help you determine if something is wrong with your cells or if they are nearing death.

- The cell discharges very fast.
- The battery gets very hot when you are charging or using it.
- You have been using the battery frequently for over two years.
- The battery's capacity has dropped to below 80% of its original capacity
- It takes an unusually long time to recharge
- There are cracks, dent, or deformities on the cell.

If you notice any of these indications, cease use to prevent personal risk or harm. Below are some suggestions that can help you get the most life from your 18650 Li-ion cells.

1. Always monitor the temperature of your batteries. If they run too hot too often, they will age prematurely.
2. To get the most from your battery, always keep its voltage between 3.0V to 4.0V. This will take some effort as most cells discharge to

2.5V before needing a recharge and have a voltage of 4.2V when fully charged. In general, pushing a battery to its limits will stress the battery. However, maintaining a moderate voltage has been shown to increase the battery's lifespan.

3. You must charge your batteries using the correct charger. Also, the charger should turn off once the batteries are full, since most chargers continuously top off the battery to ensure that it is at max capacity until discharged. This practice, as mentioned previously, can cause stress and premature aging.

4. Most of the chargers available today are capable of charging at 4A up to 6A. While this might be great for the fast charge feature, rapidly charging a battery can lead to aging and put additional stress on the battery. You also don't want the charge rate to be too low. A low charge rate means that for a battery with a high capacity, it will charge slowly. And as stated before, if a battery is charged too slowly, it can begin to self-discharge while charging, thus causing the battery to be less efficient. A healthy charge rate range is from 0.1A to 3A depending on the battery size.

Important: Li-ion 18650 batteries are made and sold to be used in systems integrated with proper protection wiring or in a battery pack with a battery management system or a PCB. Using the wrong type of 18650 cells in a device or incorrectly using tone can result in fire.

Now that you are more familiar with the 18650 battery, the factors to consider when selecting one and how to care for it, let's discuss where to source batteries.

Sourcing Your Batteries

There are several ways you can source Li-ion cells for your DIY project. You can buy new ones or in the true spirit of DIY repurpose old or dead batteries. We will cover the different options available based on price and availability.

Purchasing New Batteries

According to Bloomberg NEF in 2019, of the 316 GWh of lithium available globally, 73% comes from China. From this, it is clear that the majority of the cells you will find will come from China. Because they have a big market share, you are assured of a fair price. For this project, you don't need too many cells, so you can order some online from retail sites such as Amazon or eBay. However, even with the convenience of ordering your goods online, be sure to look for a credible vendor.

Read through the reviews and feedback ratings and check out the number of years they have been in business. This will help you in knowing if the vendor is credible and also assure you of the quality of the products you are buying. You will save both time and money if you make the right vendor choice.

However, you should be careful not to buy counterfeit goods just because the price is low. With the increase of counterfeit products in the market, lithium cells, which are in high demand, have become a target for many. Counterfeit lithium cells are being sold off as knockoffs of established brands to fool unsuspecting buyers, so be wary of that. When making your DIY battery pack, it is recommended that you use quality materials because you are assured of their safety.

Counterfeit cells can be faulty or compromised, and you could get hurt if you use them. Not only are they hazardous, but they can also damage your equipment or harm you in serious ways because you do not know the materials used to make them or their load capacities. Research and do your due diligence before you make such bulk purchases.

How to Determine Whether Your 18650 Cells Are Genuine or Fake

- You can use a capacity tester and compare the capacities of two identical cells bought from different places. For instance, you can buy a cell from a hardware store you trust and compare it to the one you ordered over eBay or Amazon. If their values check out, then the battery in question is genuine. This method, however, is not foolproof because the fake cell can show that it has the same capacity as the genuine cell, but it might not be able to perform like the real one. Let's take the Samsung 25R 18650 battery, for instance. It has a capacity rating of 2500mAh and a CDR of 20A. What this means is that it can provide a current of 20A continuously until it discharges fully. So a more definitive way to test whether the cells we have are genuine would be to measure their temperatures as they discharge and compare them to the datasheet provided by Samsung, the manufacturer.
- If the findings differ from those presented in the datasheet, then the cell is not genuine. You can assume the fake cell was rewrapped and labeled

as a Samsung 25R; however, it lacks the qualities of a real 25R battery.
- You can also use the datasheet to compare other parameters such as the weight, operating, and storage temperature.

Repurposing Old Cells

Since Li-ion batteries are everywhere, it makes sense that you would think of making your battery pack from old cells. Even though they are considered dead, batteries still retain a large amount of usable energy capacity (80%). What this shows is that even though a battery doesn't have sufficient energy for its predetermined applications, it may still have enough energy to power something requiring less energy. There are many advantages to repurposing cells. For example, it is a cheap way to collect cells for your project. It is also good for the environment, as it reduces the amount of hazardous waste produced. The recycling of old rechargeable batteries had birthed a new industry and also presented the numerous possibilities of reusing old batteries instead of making new ones.

Reviving a Dead Li-Ion Cell

Lithium-ion batteries are favored for their high energy density, so before throwing your dead batteries out, follow these steps to try bringing them back to life.

1. Take the voltage reading. Li-ion batteries are known to go into sleep mode if the cell is drained too much. If your cell has a voltage of 3.7V, but it is only showing 1.5V, then it might be in sleep mode.

2. If your cell is in sleep mode, you need to wake it up with a boost by charging the battery. This method might not always work, and this should never be done using batteries that have had a voltage reading of 1.5V for over a week, as copper dendrites may have formed. In such a scenario, boosting may easily cause a short circuit. Some chargers halt the recovery process if the voltage does not increase gradually.

3. Let the battery charge for about five minutes and take another voltage reading. If the battery is good, the voltage would be higher than before. However, remember that this might not always work, so be prepared to get new batteries.

4. Charge the cell fully. Depending on the type of Li-ion battery you have, it can take three hours or more. Some chargers automatically start charging the battery once the recovery or boosting process is over. Once fully charged, use the cell in a device that is sure to stress the battery a bit.

5. Put the battery into a ziplock bag and freeze it for about a day. Ensure that no moisture enters the bag as it could get the battery wet. Then remove it and let it thaw for about eight hours to return it to room temperature.

6. Fully charge the battery. Its performance should be better, and it should also last for longer periods in between charging cycles.

7. Remember that lithium is a volatile substance. The utmost care should be taken when handling these cells.

8. There is an end to the life of a cell. This happens once all its resources are spent, and this process is irreversible. Also, recycled cells are not suitable for all applications. For instance, you can use a recycled Li-ion car battery to power the lights in your home, but you cannot use it in energy storage systems. For this purpose, you would require a new battery.

Before you can use these batteries, you must ascertain that they are in good enough condition to work by determining their state of charge. This is the level of charge a cell has in relation to its capacity. You must first determine the quality of the cell if you cannot tell its age. This will help you better understand the condition of the cell, its capacity, energy density, and also gauge if it is safe to use. To measure the quality of the unknown cell, purchase a new cell similar to the 18650 cells you already have and label it. Proceed to take the voltages, discharge rates, internal temperature, and other battery parameters. The results will help you determine the state of your unknown cell. You must inspect the cell for any signs of physical damage. Look for scratches, dents, and even tears, no matter how small they appear. If the cell is physically compromised, using it can put you in danger. A Li-ion capacity tester can also help you determine the health status of a recovered cell.

Battery Management System

A battery management system (BMS) is a system that manages the electronics of a rechargeable battery, whether it is a cell or a battery pack. It safeguards both the user and the battery by ensuring that the cell is operating within its safe operating parameters. The BMS monitors the state of health of the battery (SoH), collects data, controls environmental factors that affect the cell, and balances it.

A battery pack with a BMS connected to an external communication data transfer system or a data bus is referred to as a smart battery pack. They include additional features and functions such as fuel gauge integration, smart bus communication protocols such as GPIO options, cell balancing, wireless charging, imbedded battery chargers, and protection circuitry, all aimed at providing information about the battery's power status. This information can help the device conserve power intelligently. A smart battery pack can manage its own charging, generate error reports, detect and notify the device of any low-charge condition, and predict how long the battery will last or its remaining run-time. It also provides information about the current, voltage, and temperature of the cell and continuously self-correct any errors to maintain its prediction accuracy.

They are usually designed for use in portable devices such as laptops and have embedded electronics that improve the reliability, safety, lifespan, and functionality of the battery. These features enable the development of end products that are user-friendly and more reliable. For instance, with embedded chargers, batteries can have longer life cycles as the chargers charge the batteries to optimal, ideal specifications within the temperature limits. Accurate fuel gauges allow users to confidently discharge batteries to their limits and not worry about damaging the cell. GPIO stands for General Purpose Input/Output, and it is an interface used to connect electronic devices and microcontrollers such as diodes, sensors, displays, and so forth.

A BMS has the following functions:

1. Monitor Battery Parameters

This is the main function of a BMS. It monitors the state of a cell as represented by various parameters such as:

- Voltage — It can indicate a cell's total voltage, the battery's combined voltage, max, and min cell voltages, and so on.
- Temperature — It displays the average cell temperature, coolant intake and output temperatures, and the overall battery temperature.
- The state of charge of the cell to show the battery's charge level.

- The cell's state of health — This shows the remaining battery capacity as a percentage of the original capacity.
- The cell's state of power — This shows the amount of power available for a certain duration given the current usage, temperature, and other factors.
- The cell's state of safety — This is determined by keeping a collective eye on all the parameters and determining if using the cell poses any danger.
- The flow of coolant and its speed.
- The flow of current into and out of the cell.

2. Manage the Battery's Thermal Temperatures

Temperature is the biggest factor affecting a battery. The battery thermal management system keeps an eye on and controls the temperature of the battery. They can either be passive or active, and the cooling medium can either be a non-corrosive liquid, air, or some form of phase change. Using air as a coolant is the simplest way to control battery temperatures. These air-cooling systems are often passive as they rely on the convection of the surrounding air or use a fan to induce airflow. However, its main drawback is the inefficiency of the system. A lot of power must be used to run the cooling system as compared to the liquid-based one. Also, in larger systems such as car batteries, the additional components needed for air-based systems such as filters can increase the weight of the car, thus affecting the battery's efficiency.

Liquid-cooled systems have a higher cooling potential than air because they are more thermally conductive than air. The batteries are submerged in coolant, or the coolant can freely flow into the BMS without affecting the battery. However, this indirect form of thermal cooling can create large temperature differences across the BMS due to the length of the cooling channels. But it can be reduced by pumping the coolant faster, so a tradeoff is created between the pumping speed and thermal consistency.

3. Make Important Calculations

A BMS calculates various battery values based on parameters such as maximum charge and discharge current to help determine the charge and the discharge current limits of the cell.

- The energy in kilowatts per hour (kWh) delivered since the last charge cycle.
- The internal impedance of a battery to figure out the cell's open-circuit voltage.
- Charge in ampere per hour (Ah) delivered or contained in a cell. This feature is called the Coulomb counter. It helps determine the efficiency of a cell.
- Total energy delivered and operating time since the battery started being used
- The total number of charging-discharging cycles the battery has gone through.

4. Facilitate Internal and External Communications

BMSs have controllers that communicate internally with the hardware at a cellular level and externally with the connected devices. These external communications differ in complexity, depending on the connected device. This communication is often through a centralized controller, and it can be done using several methods. These include:

- Using different types of serial communications.
- CAN bus communicators which are often used in vehicles.
- DC-BUS communications which are serial communications over power lines.
- Various types of wireless communications.

Only higher-level voltage BMS have internal communication; low-level centralized ones simply measure cell voltage by resistance divide. Distributed or modular BMSs must utilize a low-level internal cell controller for modular architecture or implement controller-to-controller communication for a distributed architecture. However, these communications are hard, especially in high voltage systems, due to the voltage shift between cells. What this means is that the ground signal in one cell might be hundreds of volts higher than that of the next cell.

This issue can be solved using software protocols or using hardware communication for volt shifting systems. There are two ways of hardware communication, namely using an optical-isolator or wireless communication. Another reason hampering internal communications is the restriction of the maximum number of cells that can be used in a specific BMS architectural layout. For instance, for modular hardware, the max number of nodes is 255. Another restriction affecting high voltage systems is the seeking time of all cells, which limits bus speeds and causes loss of some hardware options.

5. Protection

A BMS can prevent a cell from operating outside of its predefined safe limits by using a protection circuit.

Protection Circuits

Because they have high energy densities, there are potential risks that come with using lithium or nickel-based batteries. This is why certain safety requirements and protocols were made to reduce the global risk of transporting and operating these batteries. A basic safety device used in a battery is a fuse that breaks in high current flows. Unfortunately, some of these fuses can render the battery useless; others are more forgiving and reset once the risk has passed. An essential component of every lithium battery pack is the safety protection from overheating. We have seen so far that a rise in temperature can have drastic effects on a lithium-ion battery. Gassing is another major issue faced by Li-ion cells; hence, they have vents to release gases produced in the cell. These vents are resealable, but if the internal cell pressure becomes too much because gases aren't escaping fast enough, venting with flames can occur. This can cause the top of the cell to pop off. In such a case, the battery is said to be in disintegration, and it should be left to burn out in a safe place. It is important to provide physical protection to ensure the safe operation of your end product, which, in this case, is your DIY lithium battery pack.

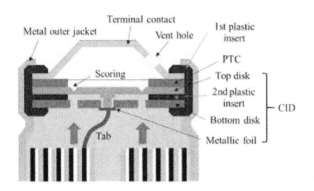

This illustration shows the basic safety features in a typical Li-ion cell.

Additional protection can be added by using protection circuits. There are two types of protection circuits, namely the primary and secondary circuits. Primary safety circuits manage functions such as protection from overvoltage cage overcurrents and changes in temperature. Lithium-ion batteries can also have secondary safety circuits that protect the cell in case the primary safety circuits fail. The protection circuits are contained in the protection circuit module, which is a printed circuit board (PCB)that contains all the components needed to make up the protection circuit. The protection circuit is part of the battery management system which manages the electronics of a rechargeable battery, including monitoring its state, reporting that data, balancing cells, controlling the environment, and protecting the battery from overcharge, over-discharge, and so on.

Protection circuits used for more demanding applications such as in energy storage systems or electric cars are operated using integrated circuits. Even though the PCBs have many advantages for larger equipment, they can contain millions of components. This is where integrated circuits (ICs) come in. They are tiny circuits that can fit inside a small silicon chip. Invented by Jack Kilby in 1958, their sole purpose is to boost the efficiency of electronic devices while helping reduce their size and production costs. As technology continues to evolve, these ICs have become more complex.

Typically, ICs use metal-oxide-semiconductor field-effect transistors (MOSFETs) to switch the lithium-ion cells in and out of the circuit. A MOSFET is a tool that uses the voltage to determine the conductivity of a device, in this case, the Li-ion battery. Its ability to change the conductivity of a device is used to switch or amplify electronic signals. Lithium-ion batteries of the same age can be connected in parallel to share one protection circuit. MOSFETs can be used as amplifiers because they have nearly infinite input impedance, thus allowing it to capture most of the incoming electrical signal. The main advantage of a MOSFET is that it does not require an input current to control the load current as bipolar transistors do. MOSFETs come in two forms, namely depletion and enhancement types. The depletion type transistors need a gate-source voltage to switch off while the enhancement ones require the gate to switch a device on. The enhancement MOSFETs can be used to boost the cell's efficiency. The PCB is responsible for protecting against:

a. Overcurrent — This protection is provided when the integrated circuits detect that the current being drawn is nearing the battery's upper current limit and thus interrupts the circuit to prevent the battery from being overly drained.

b. Overcharge — Lithium-ion batteries can be safely charged to a voltage of 4.1V or 4.2V, but not any higher. Overcharging can lead to battery damage and poses a higher safety hazard, increasing the risk of an explosion or fire. A protection circuit can be used to prevent overcharging a battery charge once its limit has been reached.

c. Over-discharge — As we stated previously, if too much current is drawn from a battery or it is discharged too quickly, it can cause damage to the cells. Lithium-ion batteries are considered empty when they discharge to a voltage of 2.5 volts per cell. Such a low discharge can stress the cell resulting in premature aging, or reduced capacity. Lithium-ion batteries often go into sleep mode once their current capacity falls under 80% of their rated capacity. This sleep mode can reduce a cell's voltage to about 1.5 volts, and if it stays in this state for too long, it can permanently lose its capacity making it unusable or dead. A protection circuit provides over-discharge protection by monitoring the discharge rate and the amount of charge drawn from the cell.

Protection circuit boards come in different varieties depending on the voltage, type of cells used, and their capacities. The most crucial parameters to keep in mind when choosing a PCB are the overvoltage and under-voltage thresholds. These are the voltages at which the PCB will cut off the cell from the circuit to prevent overcharging and over-discharging. The overvoltage protection threshold is the voltage range in which, if a battery is charged to a point within that range, it will perform well and have a longer life. For Li-ion batteries, if they are charged to 4.2V, they have more charge per cycle; however, if they are charged to 4.1V, their lifespan increases. So the PCB ensures that the battery is charged to voltages within this limit. The over-discharge protection threshold has a similar effect on the battery's capacity, charge, and lifespan as the overvoltage threshold has. A cell's capacity increases more if it is discharged completely before recharging it. However, this can stress the battery and cause premature aging, so the PCB monitors cell discharge rates and levels to prevent it from completely depleting its charge.

Ensure that you have the right PCB for your battery pack by checking:

- Protection circuitry resistance — This is the resistance caused by the PCB, which is hardly noticeable.
- The PCB's overcharge protection voltage.
- Its over-discharge protection voltage.
- The PCB's supply current — This is the current drawn from the battery to the power circuit board.
- Its overdrawn current detection ability.
- Its short circuit protection — This refers to the safety protocols engaged by the PCB, such as switching itself off in the event of a short circuit.

You must get the right PCB for your battery pack. If not, you will remain vulnerable to the safety risks posed by the unprotected battery pack.

6. Manage the Battery's Connection to the Load Circuit

A BMS can feature a precharging system that allows cells to safely connect to different loads, thus eliminating the high inrush current that floods load capacitors. This connection is controlled through electromagnet relays called contractors. The BMS's precharge circuit can be made of power resistors linked in series with the loads until the capacitors are fully charged. It can also be made of a switched-mode power supply linked in parallel to the load, which can help raise the load circuit voltage to levels similar to the battery voltage. This allows the contractors to close between the cell and the load circuit. The BMS can also have another circuit that checks whether the relay is already closed before pre-charging it. This can be due to welding, for instance, and performing this check prevents the occurrence of inrush currents.

7. Optimize the Battery Pack

To maximize the battery's capacity and prevent over- and undercharging, a BMS ensures that all the cells in a battery or a battery pack have the same voltage through cell balancing. This battery optimization can be done in four ways:

   a. By discharging or wasting energy from the most charged cells by connecting them to a load. A load is an electrical component or part of a circuit that consumes electrical power such as appliances. It is the opposite is a power source, which in this case, is the overcharged battery. This is how passive regulators work.

b. Instead of discharging the excess power from these cells, BMSs also achieve cell balancing by transferring the excess charge into the least charged cells in the connection.

c. They also reduce the charging current to a sufficient level that allows the less charged cells to continue charging while protecting the fully charged one from overcharging. This functionality, however, does not apply to lithium-based cells.

d. Finally, BMSs achieve cell balancing through modular charging.

BMS Topologies

BMS topology refers to the physical or logical layout of a battery management system. It defines how different components are placed and interconnected with each other. Alternatively, it can also describe how data is transferred between these components and the system. These layouts vary in complexity and performance. Simplistic passive regulators balance cells by bypassing the charging current when the cell's voltage gets to a specified level. However, the cell's voltage is not a very reliable indicator of the cell's state of charge, and in some cells such as $LiFePO_4$, it shows no indications at all.

The goal of a BMS is to make all the voltages in a battery cell balanced, but using passive regulators doesn't achieve this. Therefore, even though it can still perform the other functions of a BMS, it's cell balancing feature is less effective. Alternatively, active regulators intelligently switch the load current on and off whenever needed to achieve cell balancing. If these regulators only use the cell voltage as the balancing parameter, then they will also suffer the same constraints as the passive regulators.

There are three types of BMS topologies:

- Centralized — This layout has a single controller that is connected to the cells through many wires. They are the most economical, least expandable, and most complex layouts due to the multitude of wiring used to connect the BM to the battery pack.

- Distributed — This layout features small BMS boards that are installed on each cell with a singular communication cable that connects the battery and the controller. They are the easiest of the three topologies to install and offer a hassle-free assembly of the battery pack, but they are also the most expensive.

- Modular — This layout features a few controllers that each monitor and control a specified number of cells while communicating with other controllers to provide a comprehensive report on the state of the battery. These modular BMSs offer a compromise of the advantages and issues the other two layouts have.

You must note that the BMS requirements for batteries or battery packs to be used in portable applications such as powering electric bicycles or vehicles differ from those to be used in stationary applications such as in UPSs. These differences, especially in space and weight limitation requirements, require that the hardware and software implementations be tailor-made for the specific use. What this means is that before building the DIY battery pack, you must first determine the type of application the pack is to be used so that you can get the right BMS. For instance, BMSs in electric cars cannot work alone; they are a subsystem and must communicate with other components such as the charging infrastructure, the load, emergency shutdown, and thermal management subsystems. Therefore, for the battery to function optimally, the BMS must be properly and tightly integrated with these systems.

Chapter Summary

In this chapter:

- We looked at some of the materials you would need to build your DIY battery pack.
- We settled on using the 18650 cells and went on to discuss why they are the most suitable battery type for our DIY project.
- We also talked about what to look out for when selecting a 18650 battery, as well as some care tips.
- You learned about the different places you would source your batteries and the advantages of each method.
- You could get new batteries or repurpose old ones. With the latter, you would have to perform some checks to ensure that they are in good condition to be used.
- Your DIY battery pack would need a BMS to help you manage its electronics.

- We defined what a BMS was and went through its various functions in-depth.
- In a BMS, some components make up the protection circuit. This circuit keeps the battery operating within its safety limits.

We will start building our battery pack in the next chapter.

Chapter Four: Building Your Lithium-Ion DIY Battery Pack

We have finally reached the reason you are reading this book. Let's begin making our battery pack. The previous chapters served as a foundation; now, we will apply what we have learned to create a battery pack. Let's get to it!

A battery pack is comprised of several batteries, preferably identical ones or individual cells, configured in either a series or parallel connection. Sometimes, they are even built using a combination of connection types to build a battery pack with the desired capacity, voltage, or energy density. The components of a battery pack are connected using wiring that conducts electric charge between them. Battery packs often contain different features depending on the battery type used to build them and their intended usage. Rechargeable packs such as the one we will make can contain a temperature sensor, which monitors the cell's temperature; a regulator, which keeps the voltage of individual cells below the maximum allowed charge while other cells continue charging; balancers, which transfer charge from stronger cells to weaker ones to balance the pack's overall energy output; and other components that work in conjunction with these ones to increase the pack's performance and lifespan.

A battery pack has several advantages:

- It can easily be put in or taken out of a device. This allows for multiple battery packs to power the same device, providing extended runtimes. It also frees up the device to be used continuously because you don't have to wait for it to charge.

- The flexibility in pack design and implementation allows for less costly high energy cells to be connected, creating a pack ideal for any application.

- By building battery packs from repurposed or recycled batteries, you can help reduce the amount of hazardous waste produced.

Battery packs also have a drawbacks:

- It is easier for end-users to tamper with or repair a battery pack than a sealed non-serviceable battery. While the ease of access might be considered an advantage, it poses great risks. Precautions must be taken when repairing a battery pack as there are potential chemical, electrical, and even fire risks.

Now that you know what a battery pack is and its advantages and drawbacks, let's begin.

For this project, you will need:

- 18650 cells — The number of cells you need will depend on the voltage you want.
- Pure nickel strip — It is advised to use pure nickel strips rather than the nickel-plated steel strips because pure nickel strips have a lower resistance. This translates to less heat generation and wastage, giving you more range from your battery and a longer lifespan.
- Spot welder — I would recommend using a spot welder to connect the nickel strip as doing so does not add heat to the cell. You can also solder the strips, but this can be tricky as the soldering iron can add more heat to the cell, which can cause damage.
- Fuse wire
- Epoxy glue, a digital voltmeter, scissors, Kapton non-static tape or any other heat resistant tape you can find.
- A BMS (battery management system)
- Short length of silicone wire (12-16 gauge)

- Foam padding (optional)
- Shrink Wrap or tape (optional)
- A heat gun or hair dryer (if using heat shrink tube)
- Electrical connectors
- Gloves
- Safety goggles

Battery Pack Layout and Design

For this project, we will be making a 24V battery pack using 14 cells in a 7s2p configuration. What this means is that seven cells will be connected in series and two in parallel. But you can build a battery pack of any size or capacity using the same steps simply by changing the number of cells you use and connecting them in series or parallel. You must figure out the parameters of your battery. These include the voltage, capacity, and max continuous current you require. These values will help you decide the number of cells you will need and the configurations you will use. We will discuss below how to determine these specifications.

Voltage

We have defined voltage as a measure of the electrical force within the battery. A battery with a high voltage has more electrical force and vice versa. Connecting cells in series increases the overall battery's voltage. So if you want your battery pack to have a higher voltage, connect more cells in series. Higher voltage batteries are more efficient as they can provide the same amount of power as a low voltage battery even on low current. However, they are not suitable for all devices, so you must match your battery pack's voltage to that of the device or equipment you want to use it for. We also have to factor in the voltage range over the discharge curve of the battery and the voltage drop. This is the instantaneous reduction in voltage when a load is applied to the battery. So the larger the load, the more the voltage will drop, this phenomenon is known as voltage sag. The voltage will also slowly decrease as the battery discharges when the load is applied.

If the device you want to power requires a constant voltage, you might need a voltage regulator or voltage converter. A converter takes the energy from your battery and increases or decreases it to the specified voltage. The converter is also useful if you need a higher voltage but do not want to build a bigger battery. For example, suppose we needed to supply 12v to a device, but we only have space for two cells. We could build a 2s1p and then utilize a small DC-DC converter to step up the voltage. A boost converter will draw more current from the battery to make up for the increased voltage. For most devices, you can find their electrical specifications either in the manual or printed on them. On laptops, this information is usually on the battery and the battery compartment.

Capacity

A battery's capacity is the amount of charge stored in it and is determined by the mass of active materials within the battery. It represents the maximum amount of energy that can be drawn from the battery under specified conditions. We can liken this to the size of a car's gas tank where the larger the tank, the longer the vehicle will drive, provided the vehicle is in a good state. The capacity of a lithium-ion battery is represented as either watt-hours (Wh) or amp-hours (Ah).

The amp hours indicate the number of amps the battery can supply in one hour while the watt-hours indicate the number of watts the battery can supply in an hour. Amp-hours give a better indication of the total energy in the battery, while watt-hours can provide a better comparison between batteries with different voltages. To calculate the watt-hours, multiply the battery's nominal voltage by the Ah rating of the pack. In this case, if our pack is 24v 10Ah, it would approximately give us 240Wh. Capacity in this context is related to the maximum continuous discharge current rating of the battery.

For your DIY lithium-ion battery pack, you must know the exact capacity your battery will have so you can use it accordingly. If you want to build a pack to charge or fly a small drone, then a 4Ah battery is sufficient. If you were looking to charge up an electric bicycle, then you would need a battery with a higher capacity of 10Ah-20Ah. Also, note that cells do not provide the entire indicated capacity even when they are new, and capacity decreases after a few hundred charge cycles; therefore, I recommend that you build a pack with a higher capacity than you need.

Maximum Continuous Current

You also have to determine the maximum continuous current your battery pack can provide. This is the maximum current at which a battery can be discharged continuously. That is, the maximum current you can steadily draw from the battery pack without damaging it or causing it to fail. The C rate determines this current.

When covering capacity, we concluded that it is better to build a battery with a higher capacity than needed. However, increased capacity also means increased current. Powering a device with a higher current than it needs can be dangerous, so you have to ensure that the pack provides enough current. This consideration is known as the factor of safety which is determined by:

Factor of safety = maximum allowable load/actual load

To put this into context, if we are trying to power a light bulb that draws 5A continuously, then the lowest discharge current rate a battery can safely have is 5A. If you were to use a 10A battery instead, the safety factor would be 2 (5A/10A). This means that the battery can supply twice as much current as the bulb requires, doubling its runtime. Another reason why we need the factor of safety is that lithium-ion battery cells can have a reduction in performance as they approach the maximum rated values.

Now that we have a better understanding of how these three basic battery concepts affect how we design our battery, we can move on to drawing the layout. There is no standardized model or design because the layout is dependent on your needs. The circuit configuration, however, can affect the layout. For instance, a 3s3p configuration yields a cube-like shape, while a 7s2p configuration creates more of a cuboid. Whatever configuration or shape you choose, it should be simple, easy to construct, and one that adequately utilizes the space. Remember, odd-shaped designs can be hard to fit, so stick to simple rectangular or cube shapes unless the device you want to power has a unique shape.

Before you begin building your battery pack, ensure that:

- Your workspace is clean and free of clutter. Since you will be working with electricity, the last thing you want is to short out your batteries or even cause a fire because a metallic object touched your battery terminals.

- Wear your safety gloves to protect yourself, especially if you have sweaty palms, and safety goggles to protect your eyes from sparks and other particles that might fly around. You can also wear long-sleeved shirts to protect your arms from getting burned by sparks.

- Remove anything metallic you might be wearing, such as jewelry.

Step 1: The Layout Drawing

We are building a 24V battery using Samsung 25R 18650 cells, which have a capacity of 2500mAh and a max 20A current discharge rating. To attain our intended voltage, we have to connect the required number of cells in series. Since Li-ion cells have a voltage of 3.6V, we need about seven cells to reach our intended voltage: 7 x 3.6 = 25.2V. If each cell has a capacity of 2500mAh and we are using the 7s2p configuration, our battery pack will have a capacity of 2.5Ah x 2 = 5Ah, or 5000mAh.

Connection Circuits: Series and Parallel Circuits

The arrangement of your cells or batteries and how they are connected can affect the overall capacity, voltage, and current output of your system. Also, these circuits all have different wiring arrangements. You must understand these circuit types before you make your cell layout. Battery cells can be connected in three ways: in series, parallel or a combination.

In series, there is only one path for the current to flow through, whereas in parallel, there are multiple paths. The most distinct characteristic of a series circuit is that the current flows in a clockwise direction from positive to negative. In a parallel circuit, connections are made by joining the same terminals together, i.e., positive to positive and negative to negative. Also, all connections made must be between the same set of electrically common nodes. These illustrations below are examples of simple series and parallel connection circuits, where R indicates resistors.

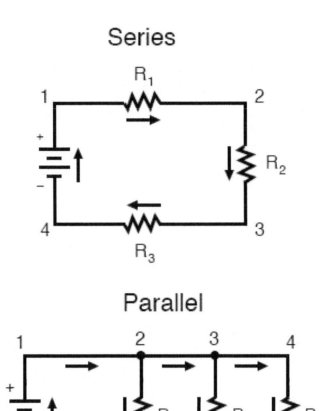

You can also connect cells using a combination of series and parallel circuits. Since series connections increase the voltage but have no effect on the capacity, and parallel connections increase the capacity but have no effect on the voltage, we can combine these two circuits to work to our advantage. This allows you to create a battery with a high voltage and capacity to match. This is the method we will use to build our battery pack.

Cell Matching

It is essential that you build your battery pack with cells that have the same voltage and capacity. Doing so can be tough, especially when using repurposed cells, but you must ensure that all cells match, which is crucial because these cells will be connected in parallel. Mismatching charge levels in a parallel circuit can lead to overheating cells and potential fires because the difference in voltage potential causes the cells with more charge to dump it into the lower charged cells generating a high current that can damage the cell. Using cells with the same voltage can counter this. Always ensure that you test each cell and note its voltage before assembling your pack. However, if you are using new cells, this step is not necessary as all cells will have the same voltage.

Now we can plan our cell configuration. This drawing will help you lay your cells correctly and also give the dimensions of your battery pack. For easy reference, I have shaded the positive cells and left the negative cells in white.

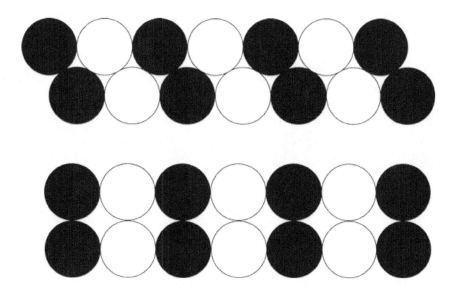

This illustration shows two layouts. The first one is more compact as the cells are tightly packed (offset), while the second one is more form efficient (linear). Each layout shows seven cells connected in series and two connected in parallel. In your layout, you might want to label the cells so that you don't get the terminals mixed up. You can add a BMS board at the end of the battery pack. Now that we are done planning the battery pack, we will prepare the individual 18650 cells. Test each cell to make sure that they are identical in voltage. If the cells are new, the voltages should range between 3.6V to 3.8V. Remember not to connect any cell whose charge is significantly lower than the others. This is why it's important to use name brand cells where you are assured of the quality.

You can also use the layout above to indicate how you will connect the circuits. Since we are using both series and parallel connections, we will connect the positive terminals of the first cells (1+), then connect those positive ends to the negative terminal of the next cell.

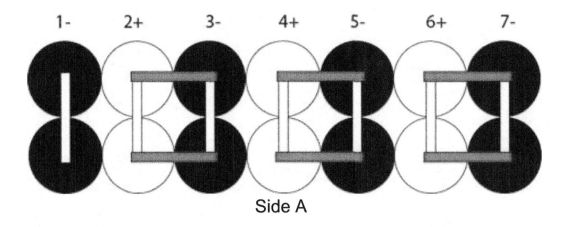

Side A

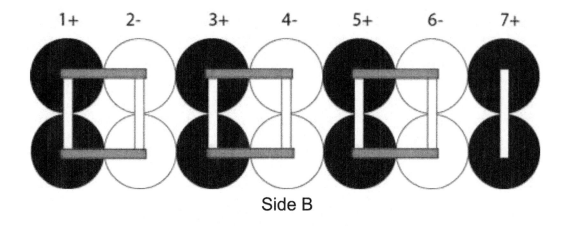

Side B

These images show how we will connect the cells.

Step 2: Assembling the Cells

Now we can assemble the cells. You have several options here. You can use glue to connect the cells as we will use for our battery pack, or you can use connectors that snap together to form your battery. Keep in mind that while connectors will align the cells for you, using them will create a larger battery than one made with glue. Here's what those spacers look like.

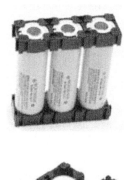

Arrange your cells as per the design and make sure that the glue dries completely. You can use the layout to double-check that everything is in the right place because any mistake can affect the battery's performance. The result should be similar to the picture below.

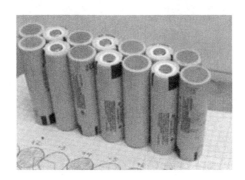

Step 3: Connecting the Cells

You can connect your cells in two ways: by using nickel strips or by using bus bars. If you don't have bolted caps, you can add the nickel strips to create the circuits. As we stated before, you will want to use pure nickel strips instead of nickel-plated steel strips because they have a lower resistance. While the nickel-plated steel ones are cheaper, they have high internal resistance, and so your battery will keep heating up and eventually become damaged. Prepare the strips before you start welding. This way, you can weld without having to stop to cut more pieces. You can get any sized nickel strip, but the thicker, the better. If you find that your strip is too fine, you can layer two or three strips together to create a connection that carries more current. The strip we use is 0.15 mm thick and 8mm wide.

To weld the piece in place, we recommend using a spot welder. This is because a spot welder allows us to securely attach the nickel strips. There are two types of spot welders, and they differ in how the electrodes are attached, namely, rigidly mounted or handheld welders. The rigidly mounted welder has two electrodes mounted on the front. It uniformly welds the strip by applying the same spring load pressure to each weld. The handheld welder has more flexibility and gives you freedom in the building process. Its electrodes are connected to the welder's main body by wires. The type of welder you use depends on the shape of your battery pack. For irregular shapes, the handheld one is ideal. No matter the type of welder you select, you are assured regardless of a robust electrical connection that is not likely to loosen up.

If you cannot access either of these soldering machines, here's how you solder the nickel strip onto the battery terminal. You will need a soldering iron and solder wire.

1. Ensure that the terminals are clean. You can use some alcohol to clean any dust that might be present on the cells. Do the same for the solder wire and clamp to remove any dust or contaminants.
2. Carefully place the nickel onto the terminals as per the layout, heat up the soldering iron and melt some of the solder onto the terminal and the strip to you want to attach. As a precaution, do not let the soldering iron come into contact with the cell any longer than necessary.
3. The soldering iron should heat up quickly. It only needs a few seconds to get hot enough to melt the solder wire. If the iron takes more than a few seconds to sufficiently heat, then something must be wrong.
4. Test to see if everything is attached securely.
5. Repeat either welding methods until all the battery cell terminals are connected.
6. Remember to take the utmost care when using such tools. If you have never used a soldering iron before, hone your

skills using dead cells.

You can also connect your cells without welding by using terminal caps. These caps are quite similar to cell spacers except that they are bolted to allow you to easily connect bus bars to a cell. They are designed like Lego blocks, with one piece fitting into another. You will need 14 terminal caps for this battery pack. Since our pack is small, we can connect all the caps first, insert the batteries, place the top layer of the caps onto the cells, and compress them or push down until they fit. Doing so does require a bit of effort, so you can use a rubber mallet to hammer them down or press them down using your own body weight.

Welding with terminal caps, however, is not suitable for larger batteries, as it can be difficult to uniformly press down on the cells. You can also damage the caps by trying to do so. With larger batteries, the column method is used to weld cells. If your battery pack configuration is 7s3p, you would first connect the caps of the three cells in parallel, since they form the columns of the battery. Once you have connected them, placed the cells, ensuring that they make contact with the ends and inserted the tip caps, similarly connect the next three cells and join these two blocks together. Repeat this process until the battery is joined as per the layout. Please pay close attention because you have to connect these cells in series. You can mark the ends so that you don't get mixed up.

Next, lay in the bus bars over the bolts as per the configuration you need. For our cell, it is as indicated in the diagrams above, and then secure them by tightly screwing the nuts back.

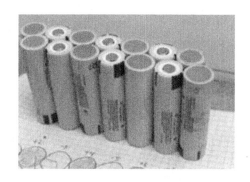

Spot welded battery pack

Battery pack with cap connectors

Ring connectors are ideal if you plan to move your battery pack around, since they are very secure. Always remember to screw the nuts back on tightly because if they become loose, it can cause a loss in power.

Now let's talk a bit about the wires used in electrical circuits. You should know which wire to use because they vary in size and conductivity. They also have capacities which indicate how much current they can conduct. This is referred to as ampacity, which is the maximum current a conductor can carry continuously while in use without exceeding safety limits.

Whenever a circuit is made, it is critical that the wiring used is properly sized for the amperage rating of the current expected to flow. The higher the current rating, the thicker the wires so that the current doesn't generate too much heat causing the wires to melt, leading to fires. Once you determine the amperage rating, you should use a wire whose gauge is appropriately matched to the battery's current output. There are many types and sizes of wires to choose from. Knowing the right wire gauge can help you make the right choice. Understanding this wire gauge rating matters, especially when you want to install a fuse wire in your battery pack. Adding a fuse to your pack is a great way to protect against short circuits, overloading, and overheating. This is because the fuse wire is designed to sense any current overloads and blow to protect the rest of the cells before the situation worsens. This wire is designed to carry the normal current provided by the battery without interrupting and to break the circuit whenever the current surpasses the set limit. Its main purpose is to limit excessive damage to the equipment, this doesn't only refer to the load but the battery pack too.

These wires are not to be confused with fuses themselves or fusible links. These wires are of a smaller gauge than the material used to connect the battery cells and are made of another material, such as copper. Fuse wires are more permanent and cannot be replaced as easily as fuses can. Thus, they are not commonly used as protection devices, but instead act as supplementary protection. They are applied in cases where installing a fuse holder is impractical.

Here are examples of some wire sizes and amp ratings.
- 14-gauge (15-amp circuits)
- 12-gauge (20-amp circuits)

- 10-gauge (30-amp circuits)
- 8-gauge (40-amp circuits)
- 6-gauge (55-amp circuits)

Step 4: Adding the BMS

Now that your cells are connected, you can add your BMS. Adding a BMS is not strictly required since we can use the battery pack as is. However, forgoing one would require that you closely monitor the pack to avoid scenarios that might cause malfunction during charging and discharging. You will also need to get a charger that can balance all the cells individually as they charge, and these can be pricey. It is easier to install a BMS and let it handle all these things unless you have a particular reason why you want to monitor the pack yourself.

As stated before, a BMS has several functions, such as monitoring battery performance parameters including the battery's state of charge, balancing the pack, protecting it from overcharging or over-discharging, and many more. Ensure that you get a BMS that matches the specifications of your battery, i.e., the voltage and the current drawn. Since we have seven cells in series, our overall voltage is 25.2V, and the current drawn is 5A, we will use a 7s BMS with a rating of 20A, which is within our battery specs. It recommended that you get a BMS that is rated higher than the current you will be drawing.

There are various BMS units available, and while no standard definitions can be used to describe all of them because they are uniquely designed, many have similar characteristics that can help you understand how to wire them. They also come with a manual to help with installation. Every BMS manufacturer has its own unique take on the design of the unit; however, there are three main connections on every unit, indicated as B-, P-, and C-. These markings show you where to connect the right wires. Now let's learn what each notation means.

1. C — This is where you connect the negative charger connector. The positive wire for your charger will go to the positive end of the battery, not the BMS.
2. B — This is where you connect the negative terminal of the battery to the BMS.
3. P — This connects to the discharge connector, i.e., this wire goes directly to the device that our battery will be powering. Please note that the discharge wire is not connected to the battery pack in any way. Instead, the current from the battery pack is routed through the BMS to the discharge connector. Think of it this way: B- is the negative terminal of the cells, while P- is the negative terminal of the pack. The positive end of the discharge connector is then linked to the positive end of the cells. The image below shows all four connections.

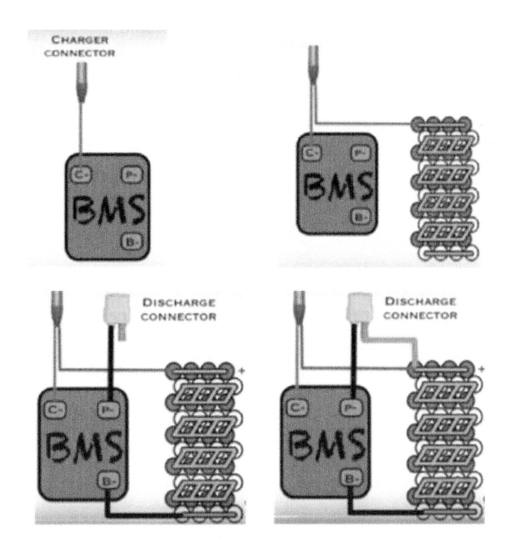

Next, you need to connect the sensor wires from the BMS to the battery. These are going to be connected to each cell group. Our BMS has seven sense wires that will connect to the seven cell groups that we have.

You can connect your BMS anywhere on the pack as long as it is secured. In our pack, we will place it on the side of the pack, place some insulation foam below it, and proceed to connect the sense wires. The foam keeps the BMS protected from any vibrations that might be generated. Now we will attach the B- wire to the negative end of the cells, as shown in the diagram above. You want to join the B- wire to batteries, so strip the wire and solder it in-between the cells so that you don't cause heat damage to any of the cell terminals. Next, connect a wire to the positive end of the battery pack that will connect to the discharge connector.

Now we connect the sense wires or the balance wires that connect to the positive ends of each cell. The BMS is labeled $B+_1$ all the way to $B+_7$ to help you connect them correctly, as shown below.

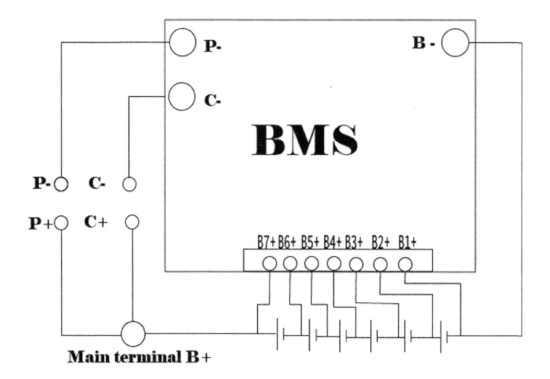

Once all the wires are connected, you can use Kapton tape (heat resistant) to secure your wires and the BMS in place. Do not forget to also cover all the terminals so that no shorts occur.

Before sealing the pack, you need to add connectors, especially if you are not using a battery case. It is through these connectors that you will connect the charger and the load. We will be using 12-gauge silicone wire for the discharge wire and 16-gauge wire for the charge wire, since it won't be carrying that much current. Start by putting on the connectors. We are using XT60 connectors for the charge wire and XT90 connectors for the discharge wires. However, you can use the same connectors for both wires, but this way, you can tell them apart. Proceed to connect the wires to the BMS as instructed above. You can also use the Anderson PowerPole connector or any other connector to match the devices or equipment you have.

Step 5: Sealing the Battery Pack

You already have a fully functional battery that can be connected to whatever device you want. You can choose to cover the battery pack or leave it be. This step is optional since it will depend on how you are using the battery. In a pack that remains mostly stationary, such as in a home energy storage unit, most people leave the battery uncovered. However, we recommend that you seal your battery as this provides extra protection from moisture and vibrations and can help preserve your battery's performance and lifespan. If your pack will be moving a lot, such as an e-bike battery, it is crucial that you seal and enclose your battery.

Lithium-ion battery packs can be sealed in two ways: using heat shrink wrap or placing it in an enclosure such as a case. Sealing the battery helps prevent it from shorting due to the exposed nickel. You don't have to use heat shrink wrap; some people use duct tape, electrical tape, plastic wrap, or other materials. We prefer using heat shrink wrap because it provides a water-resistant, but not waterproof, seal and provides evenly distributed pressure on all your battery pack components, which helps reduce the risk posed by vibrations. We also like wrapping the battery in foam before applying the shrink wrap, which protects the cells from any impact or rough treatment from accidental falls by absorbing any vibrations the pack experiences.

Now let's put on the heat shrink tube wrap. You can source them from online sites such as AliExpress, eBay, or Amazon to find the sizes you need. Please note that when quoting the sizes for your heat shrink wrap, smaller sizes refer to the diameter of the tube whereas larger sizes reference the flat width or the diameter of the circular end of the wrap. This is because as the shrink wrap sizes increase, they are no longer tubular but instead large sheets fused together at the ends. Think of it like an envelope.

Here is an easy way of calculating the size of the shrink wrap you need: Take the battery pack's height and width and add them up. The size of the shrink wrap you will need when measured by the diameter is between the sum and double that number. We double the number because unless stated otherwise, the heat shrink wrap has a ratio of 2:1. So if we require heat shrink wrap that is less than twice the perimeter size of the pack,and shrink wrap sizes are quoted in half circumference or perimeter, then the half perimeter size will be more than half of the perimeter of my pack. This will give shrink wrap that will cover the battery sufficiently.

This might be a bit complex, so let's use real values instead. Our pack is about 70mm high and about 35mm wide. The pack's half perimeter therefore is 70 + 35 = 105mm. So we will need shrink wrap whose flat width diameter is between 105 and 210mm. To be safe, a flat width of 120 to 190 will do nicely. You want it to fit as tightly as possible without damaging the battery pack components. This way, they are secure. In this scenario, a 150mm piece will do.

Something else you should note: Unless stated otherwise, heat shrink wrap only reduces 10% in size lengthwise. So when choosing a size, you want to make sure you take the longitudinal shrinkage into account.

After cutting off our 150mm piece of shrink wrap, we notice another issue. Even though the pack is well covered longitudinally, the ends are still exposed. While it may not affect the pack structurally, it does undermine the water-resistant feature of the shrink-wrap case. To resolve this, we will use a wider piece (220mm) but shorter in length to cover the longer side of the battery pack first. This ensures that the ends are sealed when you apply the second, longer wrap. If you don't have a heat gun, use a hairdryer, but note that not all hairdryers may work. The hairdryer has to run hot enough to mold the wrap. Remember not to cover your charge and discharge wires. You can also add more wrap as required to produce an aesthetically pleasing seal if that's what you want.

Here is what our pack looks like sealed using heat shrink wrap and when inside a case.

Alternatively, you can choose to place the pack in a hard case similar to those used in power banks or electric bikes. First, we will seal the battery with one layer of shrink wrap. Next, we need to connect the battery to the case. What this means is that we connect the charger negative, the discharge negative, and the positive discharge wires from the battery pack to the ones on the case, i.e., connect the C-, B-, and P- wires accordingly. Once you are done joining the wires, cover them with Kapton tape or shrink-wrap to secure everything. Then for extra protection, pack some foam padding in between the battery and the case. This will ensure that the battery pack does not move around when in use.

Before you close the case, test the battery with a voltmeter to ensure that you wired everything correctly. You can do this by taking the voltage reading of the charger and the discharge connector. Once everything checks out, you can close up the case.

Step 6: Choosing a Charger for Your Battery Pack

Now that you are done building your battery, the first thing you need to do is pat yourself on the back. You did it! The next thing on the list is choosing a charger. A good battery charger provides the base for batteries to perform well and be durable. Unfortunately, chargers are considered low priority, and as such, they are seen as an afterthought even though batteries and chargers go together. Giving the power source priority by placing it at the beginning of the project is an indicator of prudent planning. Think of the battery and charger like a car and its engine. One doesn't deliver without the other. Chargers are identified by their charging speeds with personal chargers, such as those we use to recharge our phones being low-cost versions that perform optimally when used as directed. Meanwhile, industrial-grade chargers come with special features, such as the ability to charge cells even in adverse temperatures. This is especially important since Li-ion based batteries cannot be charged in cold conditions.

Some chargers have a wake-up feature that boosts cells that are considered dead since their voltages fell below the cutoff voltage. A normal charger would regard this cell as unserviceable, meaning the pack is dead. However, a boost charger can apply a small charge to increase the voltage to anywhere between 2.2V and 2.9V per cell. This increase in voltage will activate the pack's protection circuit, initiating normal charging. A word of caution: If a Li-ion cell has had a voltage of 1.5V for over a week, dendrites may have formed, and recharging the cell may cause a short circuit compromising the cell's safety.

Lithium-ion batteries should always stay cool while charging. You should stop using a pack or a charger if its temperature rises more than ten degrees over the safety limit. Also, remember that Li-ion cells do not receive trickle charge when full or absorb overcharge. If you fully charge a battery pack but don't plan on using it immediately, you can still store it, but ensure that you use it in a week.

Charger Types

There are three main types of chargers: slow, rapid, and fast. The slow or overnight charger is the simplest type of commercial charger available. These chargers apply a fixed charge of about 0.1C, and the battery remains connected until it is full or at desired capacity. These chargers, however, have no full-charge detection mechanisms. The next type of charger is the rapid charger. These are somewhere between the fast and slow chargers in complexity and have a moderately fast charge time. When the cells are full, these chargers switch to ready mode. Many of these chargers have a temperature sensor to safely monitor the cell's thermal activity and also charge a flawed battery. Lastly, there are fast chargers that can charge a depleted battery in a short time. These short charge times are because these batteries are rated 1C.

Li-ion cells have minimal losses during charging and a high coulombic efficiency of about 99%. With a CDR of 1C, the battery will be charged to approximately 70% of its rated capacity within an hour. The rest of the time is devoted to the saturation charge, which allows the battery to achieve its full capacity. Here is a table showing the charge characteristics of Li-ion batteries.

Charge V/cell	Capacity at cut-off voltage*	Charge time	Capacity with full saturation
3.80	~40%	120 min	~65%
3.90	~60%	135 min	~75%
4.00	~70%	150 min	~80%
4.10	~80%	165 min	~90%
4.20	~85%	180 min	100%

When selecting a charger for your battery pack, there are two main factors to consider, namely the voltage and the current.

i. Voltage — You want to ensure that you choose a charger that has the appropriate voltage for your battery. To get the charger voltage, multiply the number of cells in series by 4.2V, which is the voltage of the cells when they are fully charged. In our case, there are seven cells in series, so the charger voltage for our battery pack would be: 7 x 4.2 = 29.4V. So we need a charger that outputs 29.4V. However, if you are using lithium iron phosphate (LiFePO4) cells, its voltage would be 25.9 since the cellular voltage is 3.7V.

ii. Charge current — This is the maximum current at which the battery can be charged, and it is limited by the BMS. If your BMS is rated 5A, then you wouldn't want to use a charger with a current rating higher than 5A. So it is recommended to use a charger whose current rating is well below that of your BMS. Our battery pack has a BMS rated 10A, so we don't want to exceed this limit. We are going to use a 5A charger because it gives us a good safety factor.

If you used cells that have lower charging rates or your pack doesn't have many cells in parallel, then they are the weakest link in your battery. For instance, if the cells can only be charged to 1.5A each and you only have two cells in parallel, then you can only charge 3A even if your BMS is rated for 5A. The cells in our pack have a current rating of 2.5A with two cells in parallel, so we can use a charger with a 5A rating.

You also want to consider the quality of the charger you are using. Battery chargers come in four classes.

a. Plastic charger — This is your generic plastic charger. They are sealed and the cheapest of the four types; however, they lack a cooling fan and can overheat during use. They are also more predisposed to failure and can burn themselves out, so don't leave them charging overnight as they can cause a fire.

b. Aluminum case charger with the cooling fan — This is a step up from the plastic chargers. Some of the additional features of this type include a cooling fan to help control internal temperatures, indicators, and a changeable fuse. These additional features help to make the charger more reliable.

c. Adjustable voltage chargers — These chargers are an upgrade from the aluminum case charger. What is great about them is that they have adjustable features such as allowing you to change your charge voltage so as not to charge the pack all the way up and thus increasing battery life. You can also adjust the charging current to either fast charge or slow charge your battery. You are advised to charge your battery slowly with a lower current to preserve the battery, but sometimes you want to quickly charge up your device to resume using it.

d. The Cycle Satiator — This is a high end, heavy-duty, waterproof, totally sealed, battery charger. It has a multiple current and multiple voltage feature that allows charge a battery pack with up to 8A and 60V, meaning you can charge a 14s battery pack. Its waterproof feature allows you to carry it with you without fearing any damage when wet. So you can mount this charger on an e-bike and have your battery charger everywhere you go despite the weather. You can also program your own charge profiles. This feature is especially ideal if you have several batteries. You can set a charge profile for each one of them, such as for our 24V battery, the charging current is 3.0 A with a voltage of 29.4V, 36V battery at 42V, and 4A; 48V pack at 54.5V and 4A and so on. You have the option of adding a temperature sensor so you can monitor your battery's temperature and see how hot it is getting while it charges and also set a cut off if it is getting too hot. However, it is quite pricey, and if you don't have many batteries, it is probably not worth it unless you really want the features it has. But if you have several batteries, then it is ideal for you.

The most important thing you have to remember is that your charger has to have the right voltage and amperage suitable for your battery. Here are more tips to guide you as you choose a charger.

1. A battery is charged most effectively when its charge state is low. This is because its ability to accept charge decreases significantly once the battery reaches a 70% charge or more. If a battery is fully charged, it can't convert the charge into a current, so either reduce the charge to a trickle charge or disconnect the battery.

2. Charging a battery beyond its full state of charge capacity turns the excess energy into heat and gas. This is what causes battery cell gassing and heat failure. In Li-ion cells, it can result in the deposition of unwanted materials on the electrodes. If there is prolonged overcharging the battery will get damaged permanently.

3. Use the appropriate charger for the intended battery chemistry. What this means is that if you have a Li-ion battery pack, you should use a Li-ion charger, because it is made for Li-ion batteries.

4. The Ah rating of a battery can vary slightly.

5. Charging a larger battery takes longer than recharging a smaller one.

6. Don't charge your battery using a charger whose Ah rating deviates more than 25% from your battery Ah rating.

7. A high-wattage charger can have a negative effect on your battery and cause its lifespan to shorten. Ultra-fast charging stresses the battery too much.

8. Li-ion battery chargers should have a temperature override feature that stops charging a battery once it becomes too hot.

9. Observe the battery's optimal charge temperature. This is the safest temperature they can get to while charging. Li-ion batteries should not go above 10% of its temperature when charging. You can charge the battery at room temperature, but remember that lithium-ion can't be recharged at temperatures below freezing.

Step 7: The Finishing Touches

Here you can add any last touches you want to your pack, such as a label with the battery pack details, or its specs, such as the voltage and capacity. Doing so can be very helpful, especially if you make multiple custom battery packs as it can help you remember the correct charge voltages for your various batteries.

You will also want to try out your pack and see whether it performs as it should. You can connect several halogen light bulbs to it, for instance. However, be careful that the load you connect doesn't strain the battery too much, resulting in damage. Our battery yielded an amperage of 4.9Ah on its first discharge cycle at a 0.5C rate. This means that each individual cell has a capacity of about 2.45Ah, which is about 98% of the cells rated capacity (2.5Ah). Most manufacturers often set their cells to discharge at their lowest rates, as this gives the battery its best performance. So don't be worried if you get values that are 95% of the rated capacity during real-world use. Also, the pack's capacity might increase after the first few charge cycles, especially with new cells. This is because they have now been broken in and balanced.

Chapter Summary

This chapter goes through the steps you would take to build your battery.

- We took a look at the factors that can affect the battery pack's layout and design namely the desired output voltage, battery capacity and the maximum continuous current you want the battery to discharge. Once you have these figured out, you will know the number of cells you need.
- First, you make a layout of your battery pack to identify where to place the cells and how to connect them.
- Next, you assemble your cells as per your layout and connect the cells either in a series connection, parallel connection, or a combination of both to attain your desired capacity and voltage needs.
- Then, add a BMS, which will help you monitor your battery's performance parameters. Then seal the battery or place it into a hard case.
- You also need to select an appropriate charger for your DIY pack and apply any finishing touches, such as indicating the capacity and voltage of the pack.

The next chapter will focus on the different ways you can maintain your battery pack.

Chapter Five: Maintaining Your DIY Lithium-Ion Battery Pack

Congratulations on completing your DIY battery pack. Now that you have made your battery pack, here are some of the ways you can maintain its optimal conditions so that it retains its performance and capacity.

Charging Your Pack

For most batteries, charging or discharging a battery is a chemical reaction. In Li-ion, the electrons move from the cathode to the anode and vice versa, depending on whether the battery is charging or discharging. Ideally, Li-ion batteries would last forever, but their capacities are capped by factors such as trapped electrons and parasitic reactions that can have degenerative effects on the battery components.

Most batteries are charged in a similar manner. First, let's revisit the C rate because it is the basis of the battery's usage. Batteries have a nominal capacity that's measured in Ah or mAh, which indicates the discharge current they can supply for an hour before getting completely drained. The battery pack we made has a 24V voltage and a CDR of 5Ah. This means that it can provide a maximum current of 5 amps for an hour before it discharges at a rate of 1C. If it discharges at a rate of 0.5C, then it would supply a charge of 2.5A for two hours. When charging a battery, the concept is the same. Most chargers are rated between 0.1C to 0.7C, with a 0.5 C rate being the standard.

A lithium-ion charger is mostly a voltage limiting device. Li-ion batteries differ from other batteries in that they require higher voltages per cell, tighter voltage tolerances, and no trickle or float charge when full. Since lithium-ion cannot tolerate overcharging, manufacturers have to be very strict on setting the right voltage cut-off limits as it only takes what it can absorb.

Most Li-ion cells with cathodes made from cobalt, nickel, manganese, and aluminum are charged to a voltage of 4.2V with a tolerance of +/- 50mV per cell. Some nickel-based variations can only be charged up to 4.1V while high capacity Li-ion cells going up to 4.3 volts. Increasing the voltages can increase the capacity of the cell, but this would reduce their lifespans and stress the cell posing a safety risk. The protection circuits in the BMS don't allow the charger to exceed the set voltage limit. Charging it to a lower voltage such as 3.6V can increase the battery's charge cycles, but at the cost of its run times.

A lithium-ion charge cycle has four main stages:

i. Constant current charge stage — In this stage, the battery is supplied with the maximum current the charger can output as per the C rate until the cell's voltage reaches 4.2V. At this point, the battery's charge is approximately 70% to 80% of its maximum.

ii. Saturation charge — After the cell attains its full voltage (4.1V - 4.2V), the charger becomes a voltage limiting power supply by switching to a constant voltage to eliminate any chances of overcharging the battery. A good battery charger can manage the transition from constant current to constant voltage to ensure that the battery's maximum capacity is maintained without risking any damage to the cell. Even though the cell's voltage remains at 4.2V, the charging current steadily drops to somewhere between 3% to 10 % of its rated capacity. The battery is considered to be fully charged. For our DIY pack, if the charging current gradually falls from 2.5A (0.5C) to below 0.25A, then the battery is fully charged.

iii. Ready (No current) — In this stage, no current is flowing; however the voltage might begin to drop depending on the cell's self-discharge rate.

iv. Standby mode — In this stage, if the cell's voltage begins to drop, a topping charge can be applied once every 500 hours. This is dependent on the charger and the cell's self-discharge rate. A charger can apply a topping-up charge to a battery if the voltage drops to about 4.05V per cell. The charge turns off when the voltage reaches 4.2V again.

v. There is a stage that can come before the constant charge stage. If a battery's voltage falls to anywhere between 2.8V to 3V, it is considered to be dead or over-discharged. However, some batteries can still be used even in this condition. To revive them, a boost charge is applied to wake them from their sleep mode before charging begins. The battery is charged using a 0.1A charge current until the voltage reaches about 3V.

Here is an image showing these four stages.

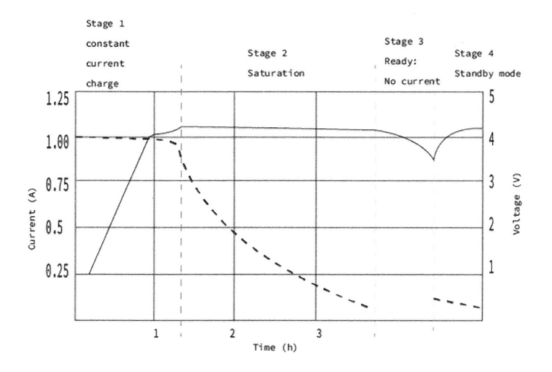

Stage 1: The voltage rises at a constant rate.

Stage 2: As it peaks, the current decreases.

Stage 3: The charging current terminates.

Stage 4: Occasional topping charge.

From the chart, we can see the voltage and current signatures of a lithium-ion battery as it passes through the various stages from constant current charge to topping charge or the standby mode. A battery is fully charged when the current reduces to about 3%-5% of the Ah rating. Our battery pack would be fully charged if the charging current fell to anywhere below 0.25A. Because of trickle charge, some chargers apply a topping up current when the cell's voltage begins to drop.

Usually, only the constant charge current and saturation stage are used since fully charging the battery can take a while. You should note that Li-ion doesn't need to be fully charged during every charge cycle, unlike other batteries such as lead-acid ones. In fact, it is not advisable to do so because the high voltage levels can stress the battery and affect its performance. Picking a lower voltage threshold and eliminating the saturation stage altogether can extend the battery's life, but at the cost of the battery's runtime (how long the battery can last on a single charge). Most chargers are set to charge the battery at maximum capacity rather than lowering the voltage to extend the battery's life.

As we mentioned before, overcharging can reduce a battery's capacity, affect its performance, and compromise its safety. If a battery is fully charged, then the ions are not moving, and most of the energy stored is converted into thermal energy. This can lead to the battery overheating resulting in gassing, explosions, or fire. Undercharging, on the other hand, doesn't pose as much risk as overcharging, but it can have an adverse effect on the capacity. For instance, if you were to undercharge your DIY battery by a mere 1%, it can cause a reduction in capacity up to 8%. That is why the chargers output voltage must be within +/-50mV of the cell's voltage, i.e., 4.1V or 4.2V. It should also have the ability to detect when a battery is fully charged and reduce the C rate from 0.5 to 0.1.

Accelerated or Fast Charging

Due to technological advances, new generation batteries have higher electron mobility resulting in faster charging times without the risk of overheating. Many manufacturers have produced various integrated circuit solutions for lithium-ion battery management systems to help achieve faster charge times. There is no conventional definition of what a fast or quick charge for a lithium-ion battery is, rather the phrase is used to refer to any charging regimen that accelerates a battery's charging in comparison to a typical 0.5C charging rate.

This accelerated charging does come at a cost. The higher the C rate, the lower the capacity. What this means is that if you charge your battery at a 0.7C rate, when the battery attains its full voltage, the capacity will be at about 70%. If you use a 0.5C rate, the battery will be fully charged when the voltage gets to 4.2V. In other words, a fast charge is ideal if you want to boost your battery, say from 25% to 60%. However, if you want a full charge, then you can attain it faster by using a lower C rate, which in this case is 0.5C. A higher C rate results in a slower top-up stage.

Another trade-off factor to keep in mind is the increase in temperature caused by fast charging a cell. Even though the thermal readings might still be within the safe operating parameters set for a particular Li-ion cell maker, they could still cause slight damage, reduce the capacity, affect the battery's performance, and reduce the number of recharge cycles the cell has. However, these effects have been mitigated by improvements made in battery technology, which make the battery cells more robust. The charging rate would have to be quite high to cause any damage.

Prolonging Li-Ion Battery Life: How and When to Charge

Even though Li-ion makes a great battery, the technology is still in development. If you take a look at the various types of lithium-based batteries available, you will notice notable improvements in increasing the battery's lifespan, capacity, voltage, and safety. Taking care of your battery can increase its lifespan and enable it to operate efficiently every time. Caring for your battery doesn't only entail conserving power. It also revolves around taking care of the physical aspects of the battery itself. As a battery-care giver, you must understand that every battery has different needs in terms of charging speed, depth of charge, loading, and exposure to extreme temperatures.

Lithium-ion batteries can age for several reasons. Ideally, its electrical charge mechanism, that is, how ions flow between the positive and negative electrodes, should work forever. However, cycling, high temperatures, and the degradation of the cell's components over time decrease its performance. This is why most manufacturers describe the life cycle of Li-ion cells anywhere between 300 and 500 cycles. However, gauging your battery's lifespan by counting the cycles can be inaccurate because discharge levels may vary; also, there is no clear definition of what a cycle is. What this means is that there are no clear definitions of the parameters that constitute a cycle. In place of cycle count, some manufacturers suggest a replacement date, but this method doesn't account for usage. A cell may fail before its replacement date due to the heavy use or the unfavorable conditions it has been put through, such as high temperatures. However, with proper use, many batteries last well past this date.

A batter's performance is gauged by measuring its capacity, which is affected by several factors such as internal cell resistance, voltage, temperature, and self-discharge rates. Shelf life can also contribute to a cell's capacity loss. Matching or balancing cells in a battery pack such as ours can also affect the overall capacity of the battery pack. Just as a shoe wears out with heavy use, so does a battery pack. The depth of discharge determines the battery's cycle count, the smaller the depth of discharge, the longer the battery can last, and vice versa. That is why you are advised to avoid full discharges and charge the battery more in-between use.

So remember that a partial discharge doesn't stress the battery pack as much so it can prolong its lifespan. The same can be said for a partial charge. You will want to avoid high currents and elevated ambient temperatures to prevent premature aging in the pack. However, for Li-ion cells, partial charging negates the advantages it has in terms of high specific energy. Elevated temperatures not only increase capacity loss but also hasten permanent capacity loss. Here is a table showing various storage temperatures and how they affect Li-ion batteries. Keep in mind that not all lithium-ion battery chemistries will behave similarly.

Temperature	40% charge	100% charge
0°C	98% (after 1 year)	94% (after 1 year)
25°C	96% (after 1 year)	80% (after 1 year)
40°C	85% (after 1 year)	65% (after 1 year)
60°C	75% (after 1 year)	60% (after 3 months)

Most lithium-ion batteries are charged to 4.2V. A reduction in peak charge voltage of about 0.10V per cell is said to almost double the cell's cycle life. For instance, a cell charged to 4.2V will have about 300 to 500 cycles, but if the charge is reduced to 4.1V, the cell's lifespan increases to about 600 to 1000 cycles. Reducing it further to 4.0V will give the cell 1200 to 2000 life cycles, and so on. The downside to this increased lifespan is that a low peak charge decreases the capacity of the cell. To put it into perspective, for every 70mV reduction in peak charge causes a 10% decrease in capacity. Not to worry, applying peak charge in subsequent cycles will restore the capacity.

A 3.9V voltage provides the optimal charge in terms of longevity. This voltage threshold eliminates battery stress, and going lower will not necessarily offer more benefits. It might cause more problems.

Dependent on what you plan to use the battery pack for, you can charge your battery either for maximum capacity or longevity. For capacity, choose a high voltage threshold such as 4.2V, and choose a low voltage threshold for longevity. Going higher than 4.2V can boost your pack's capacity, but at the risk of severely shortening service life and compromising the pack's safety.

Environmental factors other than cycling govern battery longevity as well. The most harmful thing that you can do is store a fully-charged battery in elevated temperatures. Battery packs do not die suddenly, but their runtimes gradually decrease as their capacity fades. Since low charge voltages can prolong battery life, you can make similar provisions for your battery pack. Always maintain low temperatures when charging or discharging the pack.

What Causes Lithium-Ion Batteries to Die

When lithium-ion batteries first came onto the renewable battery scene, the primary focus of many manufacturers was to maximize the cell's energy density. However, when these batteries started failing, the focus shifted to other areas, such as safety. This gave rise to Li-ion batteries that were a lot safer to use, though their capacities were consequently reduced. But even with this capacity reduction, they were still among the best performing batteries available. The issue of longevity soon came up as Li-ion batteries were now being used everywhere. Their use in applications such as powering electric cars, energy storage devices, and other heavy-duty needs has shifted the longevity factor to the forefront, focusing on why lithium-ion batteries fail. By understanding the factors that can cause batteries to fail, we are not only able to increase their lifespan, but also preserve the capacity and overall performance.

The different uses of these batteries explain the features and price tag they have. For instance, a phone or laptop battery can last up to three years or 500 cycles, while an electric vehicle battery has a lifespan of about eight years. This might seem long at first, but replacing the battery of an electric car is a lot more expensive than replacing your phone battery. If the battery can last ten to twenty years, then the longevity could justify the high price. This is why many electric vehicle battery manufacturers focus on optimizing their batteries for a long lifespan rather than to have a high specific energy. These batteries are thus bigger and heavier than normal batteries.

Let's take the batteries used in electric trains; for instance, they undergo intensive tests that strain and stress them, and their life cycles are then measured under these conditions. This way, the manufacturers can get a better read on how they will fair with daily use. The test includes charging the battery at 1.5C for less than an hour and discharging it at 2.5C for about 20 minutes at 60°C. A heavy-duty battery is expected to lose about 10% of its capacity after 500 cycles, which is equivalent to one to two years of use while maintaining a 90% capacity. However, the results show that the capacity loss is about 28%, which is a lot more than what's expected. So what would cause such an irreversible drop in capacity?

To better understand what causes such irreversible capacity losses in lithium-ion batteries, researchers performed several forensic tests, including dissecting failed batteries to find out what went wrong. They found that the lithium ions that were responsible for transferring electrons to and from the electrodes had diminished on the cathode permanently fusing with the anode. This resulted in the cathode having low lithium levels, a condition that irreversible, which caused a decrease in the battery's coulombic efficiency. Also, the components had roughened, indicating the growth of dendrites.

A battery's coulombic efficiency (CE) is defined as the completeness of the transfer of electric charge in an electrochemical system during charging and discharging. The higher the CE, the less stress there is on the battery, and thus, it should last longer. In Chapter One, we took a look at how lithium-ion batteries work. During charging, lithium atoms move from the cathode to the anode. Discharging the battery does not remove all the lithium, so the battery doesn't reset fully. A thin lithium layer then forms on the anode, referred to as the solid electrolyte interface (SEI). This layer is made up of lithium oxide and lithium carbonate, and it grows with every charge cycle. Eventually, it will become too thick and form a barrier between the graphite anode and the other battery components. A restrictive layer also develops on the cathode due to a process known as electrolyte oxidation. This oxidation can be caused by charging a battery with a high voltage, such as 4.1V per cell at elevated temperatures. The longer the battery is charged with high voltages, the faster the degradation occurs.

This sudden capacity loss can be hard to predict by testing the duration of a cell through cycling alone. Measuring the coulombic efficiency of the cell can give you more definitive results and help you verify these effects. It is stated that to achieve longevity, cells must be charged to 3.9V per cell or even lower. However, an interesting discovery was made. While charging the battery to 4.1V can cause electrolyte oxidation, and the battery decomposes faster, a battery charged to lower voltages such as 3.9V also had capacity losses due to buildup of the SEI layer on the anode. Still, a battery charged at lower voltages can last over eight years, after which the cell degradation caused by the buildup of the SEI layer occurs quickly.

Using CE, you can measure changes in the battery to detect capacity loss due to either SEI layer buildup or electrolyte oxidation. The results will then indicate the life expectancy of your battery. CE reading can vary depending on the ambient temperature and C rate. As the cycle time increases, self-discharge comes into the picture, further affecting CE. Electrolyte oxidation can somewhat cause self-discharge to occur.

A battery's CE can also be affected by the additives manufacturers use to give the battery different features such as lower internal resistance by decreasing corrosion inside the cell, reducing gassing, improving low and high-temperature performance, and so on. They make up about 10% of the electrolyte used in a cell and are often used up when the SEI layer is forming. For instance, adding about 2% of vinylene carbonate can improve the formation of the SEI layer on the anode while reducing the rate at which it forms after that. It also limits electrolyte oxidation, thus improving the CE readings.

We can thus conclude that using CE readings can help us better access the state of our battery and discover possible causes of interference in a matter of weeks instead of the alternative of waiting years for symptoms to show.

Here are some of the reasons why Li-ion batteries die.

1. Physical degradation of the electrodes and loss of stack pressure, especially in pouch style batteries. This issue can be resolved by designing the cell better and adding electrolyte additives to reduce parasitic reactions.

2. The growth of the SEI layer completely covering the anode, thus forming a barrier that hinders the graphite's interaction with the lithium ions and increasing internal resistance. This is seen as the main reason for capacity loss in many lithium-graphite systems whose voltages are kept below 3.9V. Adding additives does seem to help reduce this effect.

3. Electrolyte oxidation, which occurs at the cathode, can also lead to irreversible capacity loss. Keeping the cells at high voltages and elevated temperatures enhances this phenomenon resulting in greater capacity loss.

4. High C rates can cause lithium plating on the anode, which can also cause irreversible capacity loss and eventual battery death.

Restoring Battery Packs

If your battery pack fails, there are a few things you can do before disposing of them. As we stated previously, lithium-ion cells can go into a sleep mode that makes them appear to be dead. This happens once the capacity of each cell goes below the specified level and results in the battery cutting off its current supply as a way of preventing irreversible damage to the cell. That is why you must first try reviving dead Li-ion cells in case they are in sleep mode to wake them.

Here's what you need to do to revive a dead battery:

1. Reading the voltage — Begin by removing the battery from its connected device. Using a voltmeter, take the voltage reading of the battery. For example, if you have a cell with a 3.9V when fully charged and the volt reading you currently have is 1.5V, then the battery is likely in sleep mode.

2. Use the appropriate charger — Some battery chargers have analyzers that can either boost, recover, or wake up a battery. However, this does not always work, and you should never try to boost a cell whose voltage has been below 1.5V for more than a week. Though a charger can potentially revive a cell, doing so is dangerous. If you must use this method, take extra care to insert the cells properly into the charger.

3. Alternatively, you can boost the dead cell's voltage using a healthy cell. You can do this by connecting the cells in parallel and let it charge for approximately 30 seconds.

The battery's internal temperature can rise quickly, so do not connect the cells any longer than that. Then using a voltmeter, check its voltage. The cell's voltage should have risen to about 3.6V. You can try placing it in the charger to see if it will charge.

4. Re-read the voltage — Let the battery charge for a few minutes using the charger with the wake-up feature, and then take its voltage again. Alternatively, you can check the manual to see how long it should take. Reviving an old lithium battery may not be sufficient enough to make it usable, so you might need to buy a new battery.

5. Charge and discharge the battery — Return the battery to the charger and let it charge fully. This might take about three hours, depending on the type of lithium-ion battery and charger you have. Some chargers automatically move from recovery mode to charging so you can leave the battery in. Once it is fully charged, discharge it using a heavy load.

6. Freezing the battery — This point might seem misplaced, but bear with me; it may help you revive your dead lithium-ion battery. After discharging the battery, seal it in an airtight and waterproof bag, then place it in the freezer for a day. Make sure that water and moisture cannot enter the bag because they can damage the battery. After 24 hours, take the battery out of the freezer and let it thaw to bring it back to room temperature. This can take up to eight hours.

7. Recharge the battery — After the battery is back to room temperature, ensure that it is dry and recharge it fully using your Li-ion charger. Hopefully, this step will revive the battery and let it hold store charge again, increase its performance, and last longer. If these steps fail, then you need to get a new battery.

Maintenance Tips

1. Keep the battery pack at a moderate operating and storage temperature — Think of it this way: Food in the refrigerator stays fresh longer due to lower temperatures. Similarly, cool air preserves the battery. These moderately low temperatures prevent parasitic reactions between the electrolyte and electrodes, such as the growth of dendrites, from occurring. For lithium-ion batteries, a combination of high voltage and elevated temperature can be catastrophic. So always keep your battery pack in a cool, dry place at a charge state of approximately 50%.

The battery must stay cool or get slightly warm when charging; a higher thermal reading could indicate a problem. Don't charge the pack if the ambient temperature is above 50°C.

2. Avoid deep cycling — The cell wears down during every charge-discharge cycle. Hence, it is recommended to partially discharge your battery rather than having a full discharge. Only fully discharge your battery if you need to recalibrate it, as is the case with smart batteries. You can turn your battery into a smart battery by adding a fuel gauge to it. In this case, a full discharge can be applied when the gauge becomes inaccurate every one to three months.

Li-ion batteries last longer when operated between 30% and 80% of their rated state of capacity. Disrupting the charge cycle for a lithium-ion battery will not cause any harm to the cells. You also don't want the battery to have a low charge too often as it can turn off the protection circuit.

3. Avoid abusing the battery — Abuse, in this case, can mean two things—misusing the battery to power

equipment or devices it cannot without stressing the battery, or physically damaging the battery pack by either dropping or mishandling it. Harsh discharges and rapid recharging can put a lot of strain on a battery, so ensure that you use your battery for the intended application. The occasional rapid discharge might not cause a lot of damage to your cells, but using it for the intended purpose reduces any load-related stress that might affect the battery.

4. Avoid ultra-fast charging — We already went through how accelerated charge rates can affect Li-ion batteries, so to keep your battery pack working, charge your pack at the recommended charge rate.

5. For new batteries, applying a topping charge is enough since no priming is required for lithium-ion cells.

6. While you can charge your battery with a load connected to it, a parasitic load can affect full charge detection causing the battery to get overcharged or cause it to charge in mini-cycles.

Chapter Summary

This chapter covered battery pack maintenance, which is a vital aspect of any DIY battery making project.

- We looked at what charging a Li-ion battery entails.
- A Li-ion battery has four main charging stages:
  - The constant current charge, where the voltage rises at a steady rate.
  - The saturation stage, where the current decreases as the voltage peaks.

- The ready stage, where the charging current terminates because the cell has reached its maximum voltage.
- The standby mode, where an occasional topping charge current is applied to boost the cell's voltage when it starts self-discharging.
- Some chargers have a feature that allows them to revive batteries whose voltages have dipped to 1.5V. In such cases, when a cell is connected, it will be precharged to wake before being charged up to capacity.

- We also discussed the various ways you can prolong your battery pack and the ideal time to charge a battery.
- You learned some of the reasons why batteries fail or die, how to restore them, and restated those maintenance tips.

Final Words

As technology advances, our energy needs evolve as well. New devices are being made that require versatile power options that have more capacity, higher voltage, or higher current. As we have seen, lithium and lithium-ion batteries can provide all these features. From high energy density to low self-discharge rates, high voltage, and cell capacity, among others, Li-ion batteries have quickly surpassed other battery chemistries such as lead-acid and nickel-based ones to become the go-to power source for most portable devices.

Do you love working with your hands? Would you rather make things yourself? Do you have a device or project that needs a power source, and it requires lithium-ion batteries? If you answered yes, then this book is for you.

This book offers a thorough guide on how to build a DIY lithium-ion battery that's comprehensively detailed so that even someone with little to no prior experience can build a pack. We detail alternative methods you can use whenever there is something that requires specialized skills, such as soldering. In this case, you can weld or use connectors. The book is written in an engaging and easy-to-read format that allows you to easily jump to the section you want. If you are interested in battery maintenance, you can skip to Chapter Five, but if you want to go straight to where we start building the battery, go to Chapter Four. However, we recommend reading the whole book so that you have a full understanding of the situation.

We have covered a lot throughout this book, but let's review some of the major points.

In Chapter One, we covered lithium-ion battery basics such as how they work, the different forms they come in, and the various types available. We also looked at some of the terminologies, terms, and definitions used in relation to batteries. This chapter lays the groundwork, so you understand what we mean moving forward. Chapter Two took a look at the various applications of Li-ion batteries, such as battery energy storage systems, uninterrupted power supply, powering electric vehicles, power trains, and so on. Even though Li-ion is pretty great, it has some drawbacks, which necessitate that safety measures are taken when handling, using, and disposing of lithium-ion cells. We also covered reasons why Li-ion cells can fail, such as from bad design, misuse, aging, or external factors like temperature.

In the third chapter, we gathered the materials we will need for our DIY battery pack. We discussed cell ratings and how they can affect our decision of which cell to choose. We settled on using the 18650 cells since it is readily available. We looked at what an 18650 cell is, the factors to consider when selecting it, and how to care for your cell. We also looked at the different places you can source your cells, the BMS, and why you should have one in your pack.

Chapter Four is where we began building our battery pack. After discussing how voltage, capacity, and CDR can affect the pack's layout and overall design, we followed the seven steps listed to create the battery pack. Each step is broken down so you can better follow what is going on. After making your battery pack, you need to know how to charge, maintain, and restore it. This is what the last chapter is about. We talk about the different reasons why Li-ion batteries die and ways you can prolong your battery's lifespan.

The most valuable takeaway is that you learned how to make your very own DIY battery pack. By following the steps in this book, you can safely replicate the process and build any size battery you want. So you can get back to that project that stalled because you couldn't find a battery that suited your needs or could fit into the space you have. Now you can build a custom battery pack of any capacity you want.

Happy DIYing!

Printed in Great Britain
by Amazon